Time & the Instant
essays in the physics and phil

Time & the Instant

essays in the physics and philosopy of time

Edited by
Robin Durie

CLINAMEN PRESS

Collection © Clinamen Press 2000
Editorial selection and control © Robin Durie 2000
Individual essays © individual contributors 2000
Translation of Chapter 4 © Mary McAllester Jones 2000

Clinamen Press Limited
Enterprise House
Whitworth Street West
Manchester M1 5WG

www.clinamen.co.uk

Chapter 4 published in French by
Éditions Stock as Chapter 1 of *L'Intuition de l'instant*
© Éditions Stock, 1931
Éditions Stock, 23 rue de Sommerard, 75005 Paris

Chapter 3, 'Memory of the Present', is included with the kind agreement of Athlone Press, and is to appear in an anthology of Bergson's work, Athlone 2001.

All rights reserved. No part of this edition may be reproduced, stored in or introduced into a retrieval system, or transmitted, in any form or by any means (electronic, mechanical, photocopying, recording or otherwise) without the written permission of the publishers.

A catalogue record for this book is available from the British Library.

ISBN1 903083 22 2

1 3 5 7 9 8 6 4 2

Typeset in Simoncini by Northern Phototypesetting Co. Ltd., Bolton
Printed and bound in South Wales by Creative Print and Design, Ebbw Vale, Wales

'My mother I can recall perfectly… She spent her life making tea to pass the time and singing snatches of old songs to pass the meantime.'
Flann O'Brian, *The Third Policeman*

Contents

	Notes on contributors	ix
1	*Robin Durie*: The Strange Nature of the Instant	1
2	*Henri Poincaré*: The Measure of Time	25
3	*Henri Bergson*: Memory of the Present and False Recognition	36
4	*Gaston Bachelard*: The Instant	64
5	*Julian Barbour*: Time, Instants, Duration and Philosophy	96
6	*Lee Smolin*: The Present Moment in Quantum Cosmology: Challenges to the arguments for the elimination of time	112
7	*Keith Ansell Pearson*: Duration and Evolution: Bergson contra Dennett and Bachelard	144
8	*John G. Cramer*: The Plane of the Present and the New Transactional Paradigm of Time	177
9	*David Webb*: The Complexity of the Instant: Bachelard, Levinas, Lucretius	190
10	*Dean J. Driebe*: Time, Dynamics and Chaos	217
11	*David Wood*: Time-shelters – an essay in the poetics of time	224
	Index	242

Notes on Contributors

Keith Ansell Pearson is Professor of Philosophy at Warwick University. He has written extensively on Deleuze, Nietzsche and Bergson amongst others. He is the author of the trilogy, *Viroid Life* (1997), *Germinal Life* (1999) and *Philosophy and Virtual Life* (forthcoming), as well as *Deleuze and Philosophy* (1997), all published by Routledge.

Gaston Bachelard (1884–1962) was Professor of Philosophy at the University of Dijon, and later held the chair of History of Philosophy of Science at La Sorbonne. He exerted a decisive influence on the philosophical development of Canguilhem, Foucault, Althusser, Hyppolite and Bordieu amongst others. Of his many writings, *The Poetics of Space*, *The Psychoanalysis of Fire*, *The Poetics of Reverie*, *The Philosophy of No* and *Water and Dreams* have been translated into English. As yet, little by way of his earlier studies in the philosophy of science has been made available to the English-speaking audience. In 2000, Clinamen published Mary McAllester Jones' translation of his important early work, *The Dialectic of Duration*.

Julian Barbour has worked and published on foundational issues in physics for 35 years. He is not affiliated to any academic institution, but has funded his research through translating scientific papers from Russian and German. He is the author of *Absolute or Relative Motion?* (Cambridge University Press, 1989), and the recently published *The End of Time* (Weidenfeld & Nicolson, 1999).

Henri Bergson (1859–1941) was awarded the doctorate of letters *à l'unanimité* on the basis of his theses – supervised by Pierre Janet – on *The Sense of Place in Aristotle* and *Time and Free Will* (1889). He published *Matter and Memory* in 1896, and *An Introduction to Metaphysics* in 1903, to little general notice, but with the publication of *Creative Evolution* in

1907, he found fame with the general public. In 1922, he published his notorious work on Einstein's theory of special relativity, *Duration and Simultaneity* (recently published in a new critical edition by Clinamen). During the 1920s and '30s, Bergson's ideas fell from favour. However, in recent years his work has undergone a thorough critical reappraisal, leading to his recognition as one of the major philosophical figures of the 20th century.

John G. Cramer is Professor of Physics at the University of Washington. He has published numerous articles in the areas of the interpretation of quantum mechanics – in which he has pioneered his 'transactional interpretation' – Bose-Einstein interferometry, heavy ion physics and astrophysics. He is also the author of two novels – *Twistor* (1989) and *Einstein's Bridge* (1997), both now published by Avon Books.

Robin Durie is Senior Lecturer in Philosophy at Staffordshire University. He has published a number of articles on the philosophy of time within the phenomenological tradition. He is the Editor of the new edition of Bergson's *Duration and Simultaneity* (Clinamen, 1999).

Dean J. Driebe is a Research Scientist at The Ilya Prigogine Center for Studies in Statistical Mechanics and Complex Systems, at The University of Texas at Austin. He is also a visiting staff member of The Solvay Institutes for Physics and Chemistry in Brussels, Belgium. His research focuses on new theoretical tools for the investigation of the time-dependent dynamics of chaotic and complex systems. He is the author of *Fully Chaotic Maps and Broken Time Symmetry* (Kluwer, 1999).

Henri Poincaré (1854–1912) was possibly the last person to have had at his command the whole of contemporary maths *and* mathematical and theoretical physics. He did ground-breaking work in the psychology of mathematical creation and the philosophy of science, published nearly 500 papers on new mathematics, and wrote over 30 books – including *Science and Hypothesis* (1902) and *The Value of Science* (1904), for which, alongside his other works of popular science, he was granted membership of the literary section of the Institut, the highest honour awarded to French writers. Aside from his pioneering work in mathematics, he is perhaps best remembered for the revolution he instituted in theoretical astronomy, including his bold attempt to solve the famous 'problem of n bodies' inherited from Newton.

Lee Smolin is Professor of Physics at the Center for Gravitational

Physics and Geometry at Pennsylvania State University. He has written numerous research articles examining different issues emerging in the search for a possible unification of quantum mechanics, cosmology and general relativity. He is the author of *The Life of the Cosmos* (Weidenfeld & Nicolson, 1997)

David Webb is Senior Lecturer in Philosophy at Staffordshire University. Alongside published work on the problem of time in Aristotle and Heidegger, he has translated Vattimo and Derrida. He is the Editor of the recently published translation of Michel Serres' *The Birth of Physics* (Clinamen, 2000).

David Wood is Professor of Philosophy at Vanderbilt University. He has published widely on the philosophy of time and deconstruction, and has edited a number of collections. He is the author of *The Deconstruction of Time* (Macmillan; new edition, 2001).

1
The Strange Nature of the Instant

Robin Durie

Time has been a principal concern for philosophy throughout its history. With the work of Bergson and of such phenomenologists as Husserl and Heidegger, time came to assume a pre-eminent position in philosophy during the course of the last century. This was paralleled by a rediscovery of the problem of time in physics, prompted by Einstein's theories of special and general relativity, and by the development of quantum mechanics.

Despite being foregrounded as a matter for thought during the last hundred years, time remains notoriously difficult to conceptualise. Just as Aristotle chose to concern himself both with the question of time's very existence as well as with its nature, so today we have still to attain a position from which to make conclusive decisions regarding either of these matters.

The motivation for the present collection was to provide a space within which a number of the leading physicists and philosophers currently working on the problem of time could return to the fundamental questions of time's existence and nature. The aim was to allow the provocation of the disruptive notion of the *instant* to stimulate the creation or development of new concepts with which to confront the fundamental questions of time. In the pages that follow, the reader will encounter an extraordinary series of challenging and unsettling notions: time shelters, the complex instant, the fractal present, time capsules, the plane of the present, the breaking of time symmetry, time understood as a process which constructs the space of possible universes. As a preface to this new material, three remarkable essays (previously either unavailable or hard to come by for the English reader) dating from the early part of the century have also been included. Each of these essays fulfils the criterion of conceptual novelty in addressing the question of time – in Poincaré's essay we encounter a theoretical anticipation of the necessity of the relativity of time written some seven years before Einstein's famous paper of 1905; Bergson, in his paper on 'false recognition', proposes the remarkable notion of time 'splitting'; and in his essay on the instant, Bachelard develops a sustained provocation to our deep-seated assumption that time is continuous. Each one also exerts, to a greater or lesser degree, a profound influence on the majority of the essays that follow.

A final hope was that, in allowing physicists and philosophers to rub shoulders in a space opened by reflection on the instant, some of the extraordinary fecundity of recent thought about time in each discipline might permeate the traditional boundaries of academic faculties, further encouraging the generation of new concepts with which to think time. The privilege accorded to the notion of the instant in the collection can be traced back to a number of sources. Predominantly, it is the common resource utilised by both Barbour and Bachelard to call into question our deepest assumptions about time (even if they do not share a common understanding of the instant itself). This potential of the instant to disturb our philosophical convictions was already an important factor in Plato's work, as we shall see presently. Indeed, Aristotle's attempt to re-establish a philosophical grounding for our common sense assumptions about time has as the point of its critical focus Plato's recourse to the instant. The instant therefore provides a perspective from which to pose anew questions regarding the most appropriate fundamental concepts with which to conceive of time. The reader will find that, despite apparently disparate aims and methods, the same fundamental concerns shape the research into time and the instant undertaken by the physicists and philosophers represented in this volume: Are time and the universe essentially continuous or discrete? Is infinity or finitude the more appropriate concept with which to understand time's essence? How are novelty and creativity possible within the universe? Is it possible to establish a causal relation between the physical universe and the temporality of our phenomenology?

In the pages that follow, I will show that these questions were already determining the reflections on time of the first philosophers and scientists in ancient Greece – to outline, by way of introduction, the oldest known attempts to address the paradoxes inherent in the nature of time, from a point in history before physical philosophy and science were understood to be distinct. Moreover, I will seek to show how, in trying to establish the reality of time, a number of basic operative schemas came to determine the way in which time was understood in ancient Greece – a way of understanding which has only finally come to be challenged as a consequence of the rediscovery of time in the physics and philosophy of the last century. Indeed, it is these very operative schemas which the new concepts encountered in this collection ultimately call into question.

* * *

Today, the most audacious challenge of all to our common sense assumptions about the nature of reality has been reawakened in the name of the instant. Just as two and a half thousand years ago Parmenides had sought to demonstrate that logic allowed no room for time in the universe, so today Julian Barbour argues that the universe is timeless. There is, Barbour argues, no real time – time is just an appearance manifested by timeless instants.

Parmenides' denial of time's reality emerged from a tradition in which time already occupied a central position. As David Wood reminds us in the concluding essay of this collection, the very first fragment of Greek philosophy already talks of time. The saying of Anaximander is the expression of a representative of a group of thinkers who are to be distinguished from their predecessors, Aristotle tells us, because they forsook mystical speculation, seeking instead natural rather than supernatural causes for the world. It is for this reason that they deserve their title of *physiologi* as opposed to *theologi*.

Anaximander's words talk of the coming to be and passing away of things, a passage which, he says, occurs 'according to the ordering of time'. But Anaximander says more than this. He also explains that coming to be and passing away occur because things 'make recompense to one another for their injustice'. The earliest commentators on Anaximander immediately expressed their distaste for this blending of the ethical with the physical.

Around a century later, Parmenides' rejection of the reality of time is a direct consequence of the methodology urged upon the philosopher by the Goddess who 'leads the man who knows', the 'way' of *logos* which alone can lead to knowledge. The way of logos is constrained by the premise 'that *it is* [*esti*], and that it is impossible for it not to be.'[1] Parmenides deduces three sets of consequences from this initial premise:

> what is is uncreated and indestructible; for it is complete, immovable, and without end. It *was* not in the past, nor *will* it be; for it now *is*, all at once, a continuous one. [Fragment 8]

Parmenides' argument that what is has neither come into being, nor will pass away, is a consequence of the fact that it could only come into being from, and pass away into, what is not; but since what is not precisely is not, there can be no coming to be or passing away. Parmenides therefore concludes that what is, is *one* and *continuous*, all together and all at once. A further implication of this deduction is that what is is now, and has no past or future. Given the context within which this claim is made, the grounds for this deduction can only be that the past and the future constitute in some sense *what is not* – only the now, or present, is. The continuous eternity of the *one* renders meaningless the concepts of generation and corruption, of change, and of the past and the future. Since it would be essentially impossible to conceive of time bereft of these phenomena, we have to conclude that the *what is* of Parmenides is atemporal.

In addition, Parmenides has a further reason for rejecting a temporal interpretation of the eternal:

> Nor is it divisible, since it is all alike, and there is no more of it in one place than in another, to hinder it from holding together, nor less of it, but everything is full of what is. Therefore it is wholly continuous; for what is, is in contact with what is. [Fragment 8]

That which is undifferentiated cannot come to be differentiated: differentiation or separation is a product of what is not, and since what is not cannot be, so there can be no separation or differentiation. This provides a further ground for Parmenides' argument that neither past nor future can be. Within these lines, Parmenides defines the nature of the continuous [*suneches*] as that which is indivisible; it is indivisible because what is is all alike, and is moreover in contact with what is.[2] What is continuous is, therefore, an undifferentiated indivisibility. But since the being of time must, of necessity, be constituted of the past and future, as well as the present, then if that which is is to be temporal, it must be differentiable. What is, however, since it is continuous, and thus all alike, cannot be differentiated; therefore, once again, we must conclude that time cannot be.

Inevitably, many who came after Parmenides attempted to show the absurdity of these consequences which follow by strict reasoning from the premises that what is, is and what is not, is not.[3] All sought to reintroduce plurality, change and time into thinking. It fell to Parmenides' follower Zeno to reiterate his master's teaching, exposing the absurdity that befalls all arguments purporting to demonstrate the necessity of the many as opposed to the one, and of the reality of motion and time. Aristotle called Zeno's method 'dialectical': each of his arguments [*logoi*]

The Strange Nature of the Instant

begins from a conditional *hypothesis* which Zeno, using a similar style of reasoning to that of his predecessor Parmenides, then shows to harbour contradictory consequences.[4]

As G.E.L. Owen has shown, both the paradoxes of plurality, and the four paradoxes of motion recorded by Aristotle, share the common concern of demonstrating the difficulties involved in differentiating between, or separating out, the elements constituting a plurality.[5] The key to understanding Zeno's paradoxes is to be found in the second of the consequences which Parmenides draws from the premise that what is, is, and what is not, is not. We recall Parmenides' deduction that the One is of necessity indivisible, that because it is 'full of what is' – what is cannot consist of any part which is not – there is nothing that 'hinder[s] it from holding together', which is to say, no space or void [*apeiron*] separating out parts. 'Therefore it is wholly *continuous*; for what is, is in contact with what is' [emphasis added; Fragment 8, 22–25]. If we accept the legitimacy of this argument, then it is valid whether there are One or Many things, for even if things are discrete rather than continuous, and therefore separated from each other in some way, nevertheless, each of the things themselves must be 'full of what is', and therefore continuous. The limit case of the hypothesis that there is a plurality would be the *atomic* position which holds that each part of the plurality is infinitely small, without dimension, and hence indivisible. Zeno's paradoxes therefore seek to show on the one hand that, if the continuity of what is is understood as implying the plurality of what is, contrary to Parmenides' deduction that continuity is a quality pertaining to what is understood as a unity, then this continuity must be infinitely divisible, for 'there will always be others between any of them, and again between these yet others' (otherwise one would reach an interior limit of division, which would entail the separateness of the parts of what is, and hence that they are discrete rather than continuous). However, infinite divisibility is for this very reason an impossible process because it can never be achieved or completed. On the other hand, by a similar process, he seeks to show that the dimensionless atom is also a terminus that cannot be reached or achieved. For an infinite number of things which themselves are dimensionless will remain dimensionless, and therefore dimension can never be generated from what is without dimension. Zeno therefore concludes of the dimensionless atom: 'That which, being added to another does not make it greater, and being taken away from another does not make it less, is nothing.'[6] Over and above the arguments against plurality as such, Zeno also shows that there are dramatic consequences for those who, believing in the plurality of things, believe in the reality of motion and time, as demonstrated in the paradoxes recorded by Aristotle. The paradoxes explore the consequences which result from the hypotheses that either space and time are infinitely divisible (the 'dichotomy' and the 'Achilles') or that space and time are composed of dimensionless elements (the 'arrow' and the 'stadium'):

> Zeno's arguments concerning motion... are four in number. The first states that there is no motion because the travelling object must arrive at the halfway point before reaching the end...[7] The second, called the 'argument of Achilles', asserts that the slower runner will never be overtaken by the faster who pursues, for the latter must first reach the point from which the runner started, and in this way the slower runner must always be some distance ahead of the pursuer... What follows is that the slower runner will never be overtaken. The third argument... states that a flying arrow is stationary. This results by granting that time is composed of instants.[8] The fourth is an argument concerning two rows with an equal number of bodies all of equal length, the rows extending from the opposite ends of the stadium with the same speed; and the conclusion in this argument, so Zeno thinks, is that the half of an interval of time is equal to its double.[9] [*Physics* VI, 239^{b}10][10]

Given Parmenides' argument for the continuity of what is, the first two of Zeno's paradoxes therefore demonstrate that there can be no motion (and, as a consequence, we can infer, no time, to the extent that time is irreducibly related to motion or change) since any movement between A and B requires traversing an infinity of distances between A and B, each of which distances is itself composed of a further infinity of distances, and so forth. If, on the other hand, we wish to begin from the hypothesis that space and time are discrete, then Zeno would argue that either a discrete part is extended, and so continuous as such, and therefore subject to the demonstration of the first two paradoxes; or that it is dimensionless, atomic. But if this is the case, then the second pair of Zeno's paradoxes come into effect, showing that *motion is impossible in, or at, an instant*.

In the dialogue *Parmenides*, Plato depicts a young Socrates trying to refute Zeno's arguments by recourse to the theory of Forms. Among the many problems with which Socrates is subsequently confronted, the most striking is that of the 'dilemma of participation'. The Forms are both causes and explanations [*aitias*] of the way the world is. Things are the way they are because they participate in the Forms. Something is 'like' because it partakes in the Form of Likeness. But the Forms are only

The Strange Nature of the Instant

able to maintain their ontological status as causes if they *transcend* the world of things. The ontology of the Forms corresponds to the being described in Parmenides' poem – they are unchangingly eternal. The world of things is a world of flux, change brought about through coming to participate or ceasing to participate in Forms. With this dualist ontology, founded on the absolute separation [*chorismos*] between Forms and things, Plato is able to respect the ontology of Parmenides while simultaneously protecting the reality of change, motion and time. The question that remains, however, is what *participation* consists in.

The dilemma of participation in *Parmenides* stems from the following premise: 'Then each thing that partakes receives as its share [*metalambanein*] the Form as a whole or a part of it' [*Parmenides* 131^a]. But it appears that neither of these disjuncts is possible. For, if things participate in the whole of the form, and two or more separate things partake of the Form at the same time, then 'a Form which is one and the same will be at the same time, as a whole, in a number of things which are separate, and consequently will be separate from itself [*autou choris*]' [*Parmenides* 131^b]. But if it is separate from itself, then the Form is no longer one and the same. On the other hand, if things participate in only a part of the Form, then this entails that 'the Forms themselves must be divisible into parts, and the things which have a share in them will have a part for their share' [*Parmenides* 131^c]. But if the Forms are divisible into parts then this entails that they are no longer one.[11]

Can the apparently aporetic status of the relation between transcendence and participation be resolved? One point to be noticed is the fact that the first disjunct of the hypothesis yields a logical absurdity (that which is one and the same being separate from itself). This is further demonstrated in the first of the Hypotheses that comprise the remaining part of the dialogue, in which Plato deduces a series of negative conclusions that surpass even those formulated by Parmenides, culminating in the deduction that what is cannot bear any temporal predicates, and hence is *not in time at all*. But as such:

> it never has become or was becoming or was; nor can you say it has become now or is becoming or is; or that it will be becoming or will become or will be in the future. Now a thing can have being only in one of these ways. There is, accordingly, no way in which unity has being. Therefore unity in no sense *is*... Consequently, it cannot have a name or be spoken of, nor can there be any knowledge or perception or opinion of it. [*Parmenides* 141^e–142^a]

A significant consequence of this deduction is its disclosure that for Plato, being, and being in time, are irreducibly interwoven. This has a decisive influence on the way he comes to determine the nature of time as such.

Unlike the first disjunct, the second disjunct of the hypothesis yields a metaphysical rather than a logical difficulty, namely, that the Forms are divisible, and thus not one. Thus, the following deductions seek to examine the implications which follow from the proposition that unity is a whole of parts. A direct consequence of these deductions is an attempt to formulate the possibility for thinking the relation between the Forms and time, which is to say ultimately, between being and time.

The argument deduces in stages a series of properties that must pertain to what is conceived as a whole of parts – including motion and rest. We have just suggested that in the deductions from the hypothesis of unity's existence, the bond between time and existence occupies an axiomatic status. This comes to be confirmed when the question of whether what is, conceived as a plurality of parts, 'partakes of time [*chronou metchei*]', is raised. The answer proposed is that: 'of course it has being. But to be means nothing other than partaking of being in present time [*methexis ousias meta chronou tou parountos*], as 'was' and 'will' are being in conjunction [*koinonia*] with time past and time to come' [*Parmenides* 151^e]. However, the axiom that existence means to be *in the present* may itself be interpreted in different ways. For instance, Parmenides had written that 'It *was* not in the past, nor *will* it be; for it now *is*, all at once, a continuous one.' [Fragment 8] This implies that it is mistaken to talk of being in conjunction with past and future since past and future *are not*. Rather, what is exists in an eternal now. Such a line of thinking appears to be repeated in *Timaeus*, where Plato writes that it is wrong to transfer 'was' and 'will be' 'to eternal being. We say that it was and is and shall be; but 'is' alone really belongs to it and describes it truly.'[12] However, as Plato goes on to note, '"was" and "will be" are properly used of becoming which proceeds in time, for they are motions.' In fact, it is precisely the 'becoming' of being in time with which the current deduction from the second hypothesis is concerned. The timelessness of eternal being does not entail that time, or becoming in time, do not exist. And so the deduction proceeds to an analysis of becoming older and younger.

The time in which unity partakes as a consequence of participating in being is a time that *passes* [*poreuomenou*; *Parmenides* 152^a]. Because time itself passes, that which partakes of time passes, that is to say 'is ever becoming older than itself, since it advances according to time' [*ibid.*].

To be sure, I become older than I was previously, and correlatively, the age I was becomes younger than I am now, and this can be understood on the basis of time's passing. But in what way might it make sense to say of time that it passes? For if what is becomes older to the extent that it 'advances *according to* time', and its way of being according to time is to 'partake of being in present time', then how are we to understand the *advance* of the present? Would not Parmenides be correct to say that the present is simply continuous and one, since the present itself does not change, and therefore could not be said to advance?

An alternative perspective would be to think of that which participates in time as 'travelling' through or along a static or fixed time. Such a perspective might be represented by the image of a line punctuated by temporal points that define the nows, or instants, through which things pass. The notion of the punctual now is evoked in the deduction that what *is* is older and younger than itself in addition to *becoming* older and younger than itself. If we consider what is in the process of becoming older, then at a particular given 'now time [*nun chronon*] which is between [*metaxu*] the "was" and the "will be"', [*Parmenides* 152^b] it is older than it was. So conceived, it does 'not by-pass [*hyperbesetai*] the now'; thus, it stops *becoming* older, but rather *is* older.[13] If something is 'seized by the now,' it is no longer 'moving ahead' [*ibid*], for:

> to move ahead is to touch both the now and the after, to let go of the now and grasp the after while coming to be between [*amphoteron*] the two, namely the after and the now. [*Parmenides* 152^b]

We can infer from this passage that the nows are punctual because they have no dimension within which becoming can take place. Something that is fixed to a particular now is as static as that now is itself – in accordance with the argument of Zeno's paradox of the arrow. In order for something to become, it must move from one point to another, let go of one in order to grasp another. In so doing, the event of becoming occurs *between* the two points. We can further infer from this that the punctual nows are thus also discrete. But if the 'dynamic' perspective of time offered no explanation of how time passes, then this presentation of the static conception of time offers no explanation of that in which the time of a becoming which takes place between now and after might consist.[14]

We get an insight into how Plato might conceive of the becoming which takes place between the now and the after from the discussion of the third series of deductions which follow from the hypothesis that unity is [*Parmenides* 155^e–157^b]. It was established at 152^c that to move ahead is 'to let go of the now and grasp the after.' In the fundamental case of changing between being and not-being, the same principle is affirmed, such that becoming consists in 'getting a share of being', of 'taking' being, whereas perishing consists in 'letting go' being [*Parmenides* 156ª]. Similar claims are also made for coming to be *p*, or ceasing to be *p* and becoming *q*, what Aristotle calls change with respect to character [*Physics* V 225ª].

At this point, the matter of the time of becoming is finally addressed. The deduction within which the issue of the time of becoming is addressed explicitly trades in those qualities which the paradoxes of Zeno seek to show cannot but pertain paradoxically to the basic principles proposed by opponents of Parmenides. Thus, it is argued that unity does indeed come to be and pass away, is one and many, like and unlike, in motion and at rest. The possibility of making these assertions resides in the notion of becoming. To be sure, what does not partake of being cannot, *at the same time*, partake of being as well. Instead what is proposed is that a process of becoming occurs; a thing passes from one state to another. The temporal parameters of this passing have already been adumbrated: 'to move ahead is to touch both the now and the after, to let go of the now and grasp the after while coming to be between the two, namely the after and the now.' That is to say, if we consider two basic points of time, t_0 and t_1, then the temporality of an event of becoming consists first in its continuity with, the fact that it is in contact with, t_0; second, its continuity with t_1; and third, the 'between' of t_0 and t_1, within which the becoming occurs. If we consider an event *in* time, then this becoming would consist in the letting go of the state of affairs that pertains at t_0 and the grasping of the state of affairs that pertains at t_1.[15]

It is evident from this analysis that, if the points of time which figure within the discussion are indeed fundamental, as the text indicates, then whatever it is that is between these points cannot itself be temporal, or a part of time. So when – at what time – does change occur? With regard to the change between motion and rest: something cannot be moving and at rest at the same time; nor is there a time at which it is neither at rest nor in motion. But since a change between states requires continuity (contact) with both states, then it cannot occur in the time of motion alone, or the time of rest alone; and since there is no time at which the thing is both in motion and at rest, and no time in which it is neither in motion nor at rest, then it seems indeed that it is 'a strange thing in which it is when it is changing' [*Parmenides* 156ᵈ].[16] In fact, this strange thing is called the *instant*:

The Strange Nature of the Instant

> For the instant [*exaiphnes*] seems to signify something such that change proceeds from it into either state. There is no change from rest while resting, nor from motion while moving; but this instant, a strange nature [*physis atopos*], is something inserted between motion and rest, and it is no time at all [*en chrono oudeni ousa*]; but into it and from it what is moved changes to being at rest, and what is at rest to being moved. [*Parmenides* 156^{d-e}]

The instant, the time of change, the time of becoming, is, as we have been anticipating, 'no time at all' [*Parmenides* 156^e].

Why, precisely, does Plato describe the instant of becoming which lies between the separate nows as being of a strange nature? The Greek word translated here as strange is *atopos* – literally, without place. Why is the instant without place? Because if being is necessarily being in time, and the instant is that in which becoming takes place, then the instant, what comes to pass within it, and ultimately the very passing of time itself (to the extent that this takes place within the instant), have no being – to the extent that the instant is no time at all. But there is a sense in which this should not surprise us. For the instant inserted between separate nows has a nature which corresponds precisely to that which would define the event of participation between Forms and particulars, were transcendence to be maintained. Participation could take place neither in the realm of Forms nor in the realm of things, on pain of a violation of the principle of transcendence. Thus, Plato's recourse to the instant, and the metaphysical perplexity which it cannot help but generate, is a direct consequence of an apparently underlying desire to maintain the value of transcendence. It would seem, therefore, that Plato's account of time in the second and third deductions of the *Parmenides* is worked from within by the operative schemas of transcendence and the irreducible inter-relation of time and being, to the detriment of the development of a coherent conception of temporality.

Aristotle does not call into question the irreducible tie between time and being expressed in Plato's text. Nevertheless, his rejection of Plato's metaphysical transcendence is well-known, and is maintained in his attempt both to develop an account of the nature of time which marks a departure from the Platonic account, while also seeking to conserve the refutation of the conclusions of Zeno's paradoxes. The decisive move in Aristotle's treatise on time is his substitution of the notion of the *now* for that of the Platonic instant. By this move, Aristotle aims to exorcise both transcendence and any vestiges of the discrete from his conception of time. The key task for Aristotle is to account for time's *continuity*. His means for doing so can be discerned in his notion of potentiality.

Aristotle explicitly confronts the basic premises of the two pairs of Zeno's paradoxes of motion – that time (or space) consists in a manifold of dimensionless instants, or that it is an infinitely divisible manifold. In one sense, Aristotle affirms the paradoxical conclusion of the argument of the arrow, but this is because he wants to deny that motion can be said to take place in, or at, an instant (although he goes beyond Zeno in further arguing that rest at in instant is equally implausible). We recall Zeno's argument: If we assume that an instant is the minimal and indivisible constituent part of time, then during an instant, the arrow must be at rest. For if the arrow were indeed to move, during the event of movement it would occupy more space than its own length; but at a dimensionless instant, the arrow can only occupy a space equal to itself. Therefore, at an instant, the arrow can only be at rest. Generalise this to every instant, and it appears that there can be no possibility of movement – unless we argue as Plato does, that movement – or becoming – in fact takes place in that strange natured element that he calls the instant, which is between the moments of time and is no part of time itself. Aristotle denies the possibility of motion at an instant in the following way: let us assume that there can be motion at an instant; if this is the case, then the motion can be faster or slower. Therefore, let us imagine in the case of the faster motion that a distance AC is covered in the instant. In the case of the slower motion, the distance covered in the instant is AB. AB is a smaller part of AC. But this entails that the faster motion would have covered AB as a part of the distance AC covered in an instant, and it must therefore cover this distance in a smaller part of the instant in which it covers AC – but this entails the divisibility of the instant, contradicting our assumption. [*Physics* VI 234^{a}24ff] Equally, Aristotle goes on to argue that rest at an instant is impossible, since, utilising the general principle indicated in the previous argument that all motion entails being in different places at different times, then, in order that rest may be distinguished from motion, it must mean being in the same place at different times. Thus as a consequence, it ultimately makes no sense at all to speak of either motion or rest at any single instant [*Physics* VI 234^{a}33ff & *VIII* 239^{a}23ff].[17]

Aristotle's notion of continuity starts from the Parmenidean definition – namely that something is continuous if the 'adjacent' parts of which it is composed are in contact with one another – but refines this definition significantly, as we can see from his rejection of the principle of Zeno's first pair of paradoxes. Zeno had argued that if time or space were infinitely divisible, then traversing any distance would entail traversing the

The Strange Nature of the Instant

infinity of distances lying between, which is impossible in a finite amount of time. Aristotle responds to this paradox, first, by pointing out that Zeno conflates the infinite divisibility of his premise with a different sort of infinity in his conclusion, with the consequence that in fact there is no contradiction. This second sort of infinity is that of the aggregate, an infinity of addition. Certainly, infinite addition cannot be contained within finite limits; but a finite distance can be infinitely divided. The distance covered by the arrow therefore may be infinitely divided, but the sum of these infinite divisions is not itself infinite [*Physics* VI 233^{a}13ff]. In what sense do these infinities differ? Aristotle explains in an important but notoriously difficult passage:

> 'To be' means to be *potentially* [*dynamei*] or to be *actually* [*entelechia*]; and the infinite is either in addition or in division. It has been stated that magnitude [*megethos*] is not in actual operation infinite [i.e., there is a limit to the actual size things can be]; but it *is* infinite in division – it is not hard to refute indivisible lines – so that it remains for the infinite to be potentially [*dynamei*]. We must not take 'potentially' here in the same way as that in which, if it is possible for this to be a statue, it actually will be a statue, and suppose that there is an infinite which will be in actual operation. [*Physics III*, 206^{a}14ff]

The distinction between existing potentially and existing actually is decisive in Aristotle's philosophy. The nature of the distinction between infinities of division and aggregate is that the former is a potential infinity whereas the latter would be an actual infinity were it to be realised. But infinite aggregates can never be realised, so in fact, only potential infinites exist.

Zeno's dichotomy paradox fails because it turns on the existence of infinite aggregates. From these preceding arguments we can conclude that time is not composed of instants, but is rather continuous. A continuum is that whose constituent parts are not discrete, but rather in contact; specifically, two parts are continuous if they share one and the same common limit [*Physics* VI 227^{a}10ff]. The absence of any determinate internal limits means that the temporal continuum is infinitely divisible, but is so only *potentially*. This is because the division of the continuum will never yield an indivisible. Thus, as a dimensionless indivisible, the instant is not a part of time. It remains for Aristotle to demonstrate why time is a continuum, and in what its parts consist if they are not instants. This is the substantial work that is undertaken in the treatise on time in *Physics IV*.

Aristotle's treatise on time is defined by two lines of inquiry [*Physics IV*.10]: first, does time belong to beings or non-beings, that is to say, does time exist or not; second, what is the nature, or essence [*physis*], of time? The statement of and answer to the first question form a shell for the treatise, the bulk of the middle of the discussion being devoted to the second question. However, the issue of time's existence does not merely surround the issue of time's nature – it also essentially determines the way in which Aristotle frames his response to this latter question. This is evident from the aporias, or paradoxes, relating to time, which Aristotle, in his customary method, discusses immediately after stating the problem. The aporias pertaining to the question of time's existence relate to the fact that time appears to be made up of that which does not exist, namely the past – that which is no longer – and the future – that which is not yet [*Physics IV* 217^{b}33–218^{a}3]. Equally, the now, or the instant, is itself not a part of time – both for the reasons that we have already encountered, and also because, as Aristotle argues, a part measures a whole, and a whole, in turn, must consist of parts; but the whole of time does not consist of nows, and so a now cannot measure time [218^{a}3–8]. Finally, there is a further problem pertaining to the nature of the existence of the now, namely whether it is the same or different. If the now changes, if it is always different, then the question is begged of when it changes – for it cannot perish while it remains the same, and equally, it cannot perish in another now, for then two different nows would be simultaneous. On the other hand, if it stays the same, there would be no means of distinguishing between before and after. Thus, in addition to the problems of accounting for the continuity of time and the nature of the parts of time, Aristotle is also confronted by the problem of accounting for how time 'passes'.

When we turn to consider the nature of time, we find, according to Aristotle, that there are two key phenomena that pertain to temporality. First, as we have seen, time is continuous. Second, there seems to be a profound connection between time and motion [*kinesis*]. But time cannot be reduced to motion, for motion is always connected with that which moves, change to that which changes, whereas time is everywhere and not restricted to any particular thing [218^b]. Moreover, that which moves can move more or less quickly, more or less slowly; time, on the other hand, does not appear to have different rates of flows. Yet though time is not motion, neither is it ultimately separable from motion [219^a]. Where motion does not exist, so time also does not exist. The task for Aristotle is therefore to determine the precise relation between time and motion.

The Strange Nature of the Instant

Aristotle argues that it is only when we delimit [*horisomen*] change that we say that time has elapsed [218^{b}32]. Time and movement, or change, are perceived together [*hama*; 219^{a}3], so time must be related to movement in some way. Aristotle's claim is that magnitude [*megethos*], change and time follow [*akolouthei*] one another, and as a consequence, change and time both share the continuity [*suneches*] of magnitude. This is because all movement, or change, is 'from something to something' [*ek tinos eis ti*; 219^{b}10]. The magnitude of the from…to (its 'stretch') is continuous [*sunecheia*], that is to say, as we have seen, infinitely divisible [219^{b}11]. Since movement is also from…to, it must follow from, or in some sense be founded in, magnitude – to the extent that magnitude defines the stretch of the from…to – and as a consequence it must be continuous. Finally, since time is in proportion to movement, it too must be continuous [219^{b}13].

Having established the continuity of time based on the grounds that time follows motion, which in turn follows magnitude – the stretch of a from…to – which is continuous, Aristotle next seeks to establish the way in which time, motion and magnitude are delimited (since, as he has argued, it is only when motion is delimited that we say that time has elapsed). His argument is that the notions of 'prior' (before) and 'posterior' (after) pertain to magnitude, change and time. Prior and posterior hold for magnitude in virtue of relative position [219^{b}15] – indeed, they are constitutive of our understanding of a stretch (from the prior…to the posterior) of magnitude. A stretch similar to that holding between the prior and posterior of magnitude holds between the prior and posterior of the stretch of movement; and therefore a similar continuity of magnitude holds between prior and posterior and the from…to of movement [219^{b}16]. Thus, the magnitude of the prior and posterior holds for movement; time follows movement; so the magnitude of the prior and posterior holds also for time [219^{b}20].

From this, Aristotle moves to his conclusion that time is the number of motion – with respect to the before (prior) and after (posterior). As we have seen, time is disclosed only when movement is delimited or differentiated [*horisomen*] [219^{b}22]. Movement is differentiated by marking the prior and the posterior [219^{b}23]. We differentiate the prior and posterior of movement, and hence mark that time has elapsed, when we judge that one thing is different from another, and that some third thing is *intermediate* between them. We think of the *extremes* as different from the middle [219^{b}26]. The mind therefore judges that the 'nows' are two, and that one is prior (before the other) and one is posterior (after the other) – that is, there is a relative ordering [219^{b}28]. A unitary now would not, as we have seen Aristotle argue in rejecting Zeno's paradox of the arrow, let time appear, but what is bounded by different nows – one prior and one posterior – does disclose time [219^{b}29]. So in judging or differentiating the before and after of movement by enumerating the prior now and the posterior now, time is disclosed, precisely as what is counted of motion: time is the counting of change by enumerating nows as different.

We can see, therefore, how Aristotle's understanding of the nature of time, and its relation to motion (and magnitude) is fundamentally related to his understanding of continuity as an infinitely divisible magnitude, and his rejection of the possibility of motion at an instant. But if the now is not to be understood as an instant, if we are to reject the notion that the instant is a part of time, how are we to understand the now? In formulating his answer to this question, Aristotle conclusively departs from the strange natured instant that figures in Plato, while simultaneously defining the space for all subsequent conceptions of time.

In addition to the key elements of the continuity of magnitude (and the fact that motion, and then time, follow magnitude, and are hence also continuous), and the differentiation of the prior and the posterior in the above argument towards the definition of time as the number of motion, the third fundamental element is the judging of the prior and the posterior as *nows*. Aristotle's treatment of the now [*nun*] will attempt to show how the problem of the now as same and different can be resolved, thereby accounting for the possibility of the passing of time without recourse to the notion of an instant outside of, or between, time in which passing takes place. However, the discussion leaves apparent problems – the now will have to be conceived as both in time and not in time; both counting and counted; both identical and different.

First, then, Aristotle claims of the now that it functions both to divide and to connect time. On the one hand, the now divides time to the extent that it is like a 'point' which divides a line. As such, however, the now is not, as we have seen, a part of time [218^{a}6]. Nevertheless, we are constrained to think of the now in this way when we conceive of it as a limit [*peras*; 218^{a}24] which divides past and future [218^{a}10]. The now as a limit is not part of time, but is an attribute of time [220^{a}21]. On the other hand, the now unifies time by connecting past and future, as a point can be said to unify a line [222^{a}10]. Understood in this way, the now is not a limit dividing past from future, but the common limit shared

The Strange Nature of the Instant

between past and future, and therefore the ground of the continuity of time. As such, it is of time, but, strictly speaking, is not a delimitable part of time.

Second, Aristotle shows that the now has the characteristics both of identity and difference. On the one hand, the now is identical when it is that which it is when it is being [what it is] [*ho pote on*; 219^{b}15] – or, as Aristotle says, it is the same as a subject [219^{b}11] or substratum [219^{b}26]. Thus, when it performs its 'essential' function, either of limiting the past and the future or joining the past and the future, it remains self-identical [cf. 222^{a}15]. On the other hand, the now is different in the attributes it accepts [219^{b}11], or through what is predicated of it (e.g. 'before' or 'after') [219^{b}27]. Thus, the now is different as potential measure of time [222^{a}14], or, as he says, 'inasmuch as the point [*stigme*] in the flux of time which the now marks is changing (and so to mark it is its essential function) the now too differs perpetually' [219^{b}17].

Finally, to the extent that the now is related to the counting of change, it can be both numbered and numbering. On the one hand, when the soul judges that nows are two [219^{a}27], that is, one is prior and one is posterior, the nows are numbered [219^{b}25]. On the other hand, the now can be understood as a unit of counting [220^{a}4], and as such it is number numbering [220^{a}22].

We noted above that Aristotle's conception of time was shaped in part by his rejection of Platonic transcendence; we also noted that his refutation of the paradox of Zeno's dichotomy turned on his claim that continuity is that which is infinitely divisible *potentially*. Potentiality also provides Aristotle with one of his means for avoiding Platonic transcendence. Indeed, it is possible to see how potentiality enables Aristotle to overcome Plato's recourse to an instant that is not a part of time as the site of the passing of time. For Aristotle locates the continuity of time in its stretch, the stretch of the from…to… which follows the stretch of motion, which in turn follows the stretch of magnitude. The possibility of a stretch of time, of time's continuity, thus lies in the 'presence' of the 'to', the posterior, towards which time is flowing, 'within' the midst of the stretch, of the moment to come within the present moment.[18] But what is the nature of this presence? Clearly, it cannot be such as to obliterate the distinction between present and future that allows the marking of time's passing. The only way to understand the presence of the 'to-come' would appear to be in terms of *potentiality* – at this moment, the to… of the stretch, of which this moment is presently a part, is itself present potentially.

But if the to-come is already present in the stretch of time, albeit potentially, then how do we account for the novelty which the passing of time engenders? Doesn't the reliance on a potentially present to-come rob the stretch of magnitude, and hence also of motion and time, of that *openness*, the fundamental incompleteness or indeterminateness, which so characterizes the temporality of our experience?

This question may be posed from a different perspective: what is the origin of the now?[19] The difficulty of this problem resides in the fact that if time is experienced along with motion, and time is the number of motion, and what is counted when we count the number of motion are the 'nows which follow motion', then these nows are already temporal phenomena, as such a now is what it is through being temporal. It therefore appears as if the now, in and of itself, must be temporal, yet, at the same time, as the *foundation* of time, it cannot be temporal. The tension between these two imperatives is disclosed by the various apparently contradictory formulations of the now in Aristotle's treatise that we have noted. Among these contradictory aspects pertaining to the now, the most significant are those whereby the now both holds time together, providing the flow of time with its continuity; while, 'simultaneously', the now articulates, differentiates, time into the prior and the posterior. As Heidegger tries to gloss Aristotle's sentence [220^a5]: 'In each now the now is a different one, but still each different now is, as now, always now. The ever different nows are, *as different*, nevertheless always exactly *the same*, namely, now.'[20] The possibility of the now serving these two separate functions, of dividing and unifying, lies in the fact that the now itself is, apparently paradoxically, both the same (self-identical) and different (self-different). Aristotle's notoriously compressed expression of this state of affairs is: *to gar nun to auto ho pot en, to d'einai auto heteron* [219^b10-11]. Heidegger strives to make sense of this sentence in the following manner: 'the now is the same with respect to what it always was – that is, in each now, it is now; its *essentia*, its what, is always *the same* [*tauto*] – and nevertheless every now is, by its nature, different in each now; nowness, being-now, is always *otherness*, *being-other* (being-how or how-ness – *existentia* – *heteron*).'[21] Aristotle does not confront the question of the possibility of this simultaneous sameness and otherness of the now itself – indeed, neither does Heidegger. We believe that this is because the operative schema of potentiality does not allow for a thinking of originary difference, for a conception of temporal phenomena which are not simply self-identical – as we suggested above, the operative schema of potentiality would, on the contrary, close off the possibility of

The Strange Nature of the Instant

accounting for novelty as such.[22] Potentiality, finally, is always only ever appropriate for an account of the maintenance of the same, of phenomena as identical to themselves. The predominant concern with the issue of novelty in a number of the essays that follow is, therefore, one of the marks of the overwhelming attempt expressed in these essays to break free of the Aristotelian heritage of thinking time.

* * *

What is striking about the discourses of Plato and Aristotle on time is that, despite seeking to develop alternative theories from the perspectives of the instant, on the one hand, and of the now on the other, neither is able, ultimately, to think these fundamental concepts coherently. For both Plato and Aristotle, it would appear that the condition of possibility of time, and specifically of time's essence as *passing*, is in the end unthinkable as such. That which would render time possible – either as discrete or continuous – finally renders it impossible.

The essays in this volume run the gamut of responses to this aporetic inheritance – from Bergson's attempt to think novelty and creation at the heart of the continuity of time to Bachelard's attempt to renew a thinking of temporality from the perspective of discontinuity. As we indicated above, however, it is perhaps Julian Barbour's approach that is the most unsettling. Barbour's research in the fields of quantum gravity and quantum cosmology have led him to the conviction that:

> Einstein's general theory of relativity... is telling us that time does not exist as an independent thing and that change is indeed primary... When the fact that time has no independent existence is combined with the basic facts of quantum mechanics in the simplest possible way, the implications are startling: The quantum universe is static. Only timeless Nows exist. The quantum rules give them different probabilities. We experience the most probable Nows as individual instants of time. The appearance of motion and a flow of time are both illusions created by the very special structure of the instants that we experience.[23]

In *The End of Time*, Barbour explains the creation of time capsules – instants whose structure is such as to 'encode' the appearance of time – and hence of our experience of the passing of time. As he has written elsewhere: 'The instant is not in time: time is in the instant.' In his essay on Bergson in this volume, Barbour seeks to show that Bergson's insistence on the irreducibility of *durée* is neither necessary nor, in the final analysis, correct, if we wish to understand the appearance of time in the universe. He then proceeds to outline how his own theory of instants gives a more coherent account of the universe while containing within it the wherewithal to explain the appearance of time.

The stakes involved in Barbour's work could not be more significant. In 'The Present Moment in Quantum Cosmology', Lee Smolin argues that if we accept the premises of Barbour's position, then his conclusion that the universe is timeless is inevitable. Smolin's response is equally audacious: he questions whether we should accept Barbour's premises, and in so doing, seeks to develop an alternative account of quantum cosmology in which time comes to play an 'even more central role in the formulation of fundamental physics.' Smolin's approach to the reformulation of the role of time in physics itself entails, however, nothing less than 'a reformulation of the basic mathematical framework we use for classical and quantum theories of the universe.'

Alongside such reflections on the role of time in fundamental physics, the philosophical essays in this volume provoke an equally profound rethinking of our most basic temporal assumptions. Like Smolin, Keith Ansell Pearson argues that one of the most fundamental requirements of any theory of time is that it should be able to account for novelty in the universe. Ansell Pearson explicitly argues – from a Bergsonian perspective – that this will never be possible when thinking begins from the Aristotelian concept of potentiality. On the contrary, he argues that such Bergsonian concepts as the 'virtual' and 'tendencies' offer the only genuine way into thinking novelty. In his essay on the 'complexity' of the instant, David Webb also takes a critical glance back towards Bachelard, but where Ansell Pearson chooses to return to Bergson, Webb instead follows the inheritance of the instant through to Levinas and ultimately Serres. In a complex and dazzling argument, Webb shows that, although Levinas develops an altogether more nuanced conception of the instant than Bachelard, he is ultimately betrayed by an inability to think the very notion which lies at the heart of his own conception of the instant, namely 'deviation'. Yet deviation is precisely one possibility for thinking novelty within the atomist tradition, so Webb concludes by proposing the *clinamen* as a means for thinking novelty in a theory of time which begins with the instant, and, as a consequence, for thinking the complexity both of the instant, and of being as such.

The notion of a complex present is also found towards the end of John Cramer's paper. Cramer has worked on generating an interpretation of quantum mechanics that overcomes the paradoxes which bedevil the standard Copenhagen interpretation. As he argues, however, this 'transactional interpretation' of quantum mechanics itself harbours remarkable implications for our understanding of time. Using the plane of the present as his means for contrasting various physical theories of time, he concludes by arguing that the transactional interpretation yields a present whose complexity is such that it should be thought of not as a plane, but rather as a 'fractal-like surface'.

One of the main critical targets in Ansell Pearson's essay is the way in which Bachelard sought to undermine the Bergsonian thesis of continuity, and to argue that time's source is discrete. In his beautiful meditation on 'the poetics of time', David Wood weaves together the implications of this discrete conception of temporality with Bachelard's later notion of 'shelters' (borrowed from his *Poetics of Space*), in order to introduce his own new concept of 'time shelters'. In his essay, he suggests that the boundary of a shelter may be one place to begin to think the source of time's continuity *and* discontinuity. What Wood's essay implies is that the very opposition between continuity and discontinuity, inherited from the Greek tradition, may itself be at best secondary to a fundamental thinking of time.

The essays in this collection provide ample evidence of the rich seam of thinking about time that is to be found in the disciplines of both philosophy and physics. They also suggest that the novel ideas expressed therein share a common conceptual wellspring of shared concerns. Although, in the final analysis, philosophy and physics may be condemned to maintain their disciplinary boundaries, it may well be that time itself will ultimately undermine these boundaries.

Department of Philosophy
Staffordshire University
Stoke-on-Trent, ST4 2XW, UK.

Notes

1. Parmenides, Fragment 2. Cited according to Burnet's translation in *Early Greek Philosophy* (New York: Meridian Books 1957; 1st edition, 1892). The assertion 'it is' constitutes the simple claim that all that exists, exists. All further references will be by fragment, according to this edition.
2. As we shall see, Parmenides' follower Zeno maintains this injunction against the divisibility of what is continuous. It is only with Aristotle – who refines Parmenides' proposition that the parts of a continuum are in 'contact' with one another – that this assumption is undermined: he demonstrates that the

continuum is, in fact, infinitely divisible – potentially – but that such division never itself yields an indivisible (such as an atom).

3 See, for instance, Plato's *Parmenides*, trans. H.N. Fowler (Loeb Classical Library; Cambridge, Ma.: Harvard University Press, 1939), 127^e1–128^c6.

4 The logical form of Zeno's arguments – *reductio ad absurdum* – is employed by Parmenides; eg. if there is coming to be, what comes to be comes either from what is or what is not; but if it comes to be from what is, then it already is, and has no need of coming to be; and if it comes to be from what is not, then, as Parmenides says, 'what need could have made it arise later rather than sooner' [Fragment 8]. Therefore, there is no coming to be. One of the arguments against plurality recorded by Simplicius is as follows: 'If things are many, the same things must be both finite and infinite in number. For if things are many, they must be just as many as they are, neither more nor less. But if they are as many as they are, they will be finite in number. If things are many, they will be infinite in number. For there will always be others between any of them, and again between these yet others. So things are infinite in number. But the same things cannot be both finite and infinite in number. Therefore things are not many.' [Fragment 11; cited in H.D.P. Lee, *Zeno of Elea* (Cambridge: Cambridge University Pres, 1936.)]

5 G.E.L. Owen, 'Zeno and the Mathematicians', in R.E. Allen & D.J. Furley (eds), *Studies in Presocratic Philosophy II* (London: Routledge and Kegan Paul, 1975), 145ff.

6 Lee, *Zeno of Elea*, fragment 4.

7 The paradox thus briefly stated, the so-called 'dichotomy', turns on the fact that in order to travel a certain distance, half that distance must be covered beforehand; but in order to cover this latter distance, half of it must also be completed; and so forth, resulting in the paradoxical conclusion that, in order to cover any distance at all, it is necessary to have first covered an infinity of distances which has no first term, and therefore, the traveller never gets started.

8 The paradox of the arrow is particularly influential on the way Aristotle develops his treatise on time. If we assume that an instant is the minimal and indivisible constituent part of time, then during an instant, the arrow must be at rest, for were it to move, the instant would have to be composed of parts, contrary to our assumption, parts corresponding to the different places occupied by the arrow as it moves. But if the arrow cannot move 'during' an instant, then it must move, so to speak, in the 'time' between instants. But this entails that time is made of elements which are not time.

9 Assume that both rows are of length x. At the start, the right end of the first row touches the left end of the second row. After an instant of time, the first row has moved its length x to the right and the second row has moved its length x to the left. Now the right end of the second row touches the left end of the first row. But on the principle of the punctuality of the instant, at

no time have the midpoints of either row passed each other, despite the fact that they clearly have passed. [cf. Owen, *op cit*, 151]

10 Aristotle, *Physics I–IV*, trans. P.H. Wicksteed & F.M. Cornford (Loeb Classical Library; Cambridge, Ma.,: Harvard University Press, 1929).

11 In addition, there is a further consequence, equally troubling to the theory. If each thing which participates in the Form partakes of a part of the Form, then it follows that each thing *has its own* part of the Form. The alternative to this proposition would entail the possibility that two things participated in the same part of the Form. But this possibility would result once again in the consequence detailed in the first disjunct, namely that one and the same Form would be in separate things, and hence separate from itself. However, the consequence of arguing that a thing partaking of a Form *has its own* part of that Form is that the specific part of the Form which is thus partaken of owes its existence as a part to the thing. As Aristotle writes: 'if the Ideas are such as some hold, the substrate [*hypokeimenon*] will not be substance [*ousia*]; for the Ideas must be substances, but not involving a substrate, because if they did involve one they would exist in virtue of [*esontai*] its participation in them.' [*Metaphysics* 1031^b15] But clearly this consequence reverses that aspect of the theory of Forms which accords ontological priority to Forms, and hence bestows upon them the status of *aitia*, the originary cause or principle of things.

12 *Timaeus* 38^a; trans. R.G. Bury (Loeb Classical Library; Cambridge, Ma: Harvard Univ. Press, 1929).

13 As Owen has suggested (*op. cit.*, p. 152) there is here in Plato an anticipation of the problem which Aristotle will subsequently confront in his treatise on time, namely the apparent aporia that the now itself cannot pass – for it must do so either in itself or in another; but the former entails that the now cease to be (itself) when it is still present, whereas the latter entails that nows are discontinuous or that the now lingers into a subsequent now, thereby stretching into a period and so losing its character as now. Aristotle will seek to develop this latter suggestion, understanding the now dynamically as progressing through time. However, we have seen that Plato rejects a dynamic conception of the now. In fact, as we shall see, Plato will seek to make sense of a notion of the possibility that there is 'something' which is *between* nows.

14 We may surmise that Plato himself would not have felt the aporetic force of the implications of either the static or the dynamic conceptions of time. For, in *Timaeus*, he conceives of time as following motion, as being the image of becoming. Motion therefore does not need time *in* which to become, and so could plausibly occur between points of time – although clearly, if motion *is*, it must be in time. On the other hand, it is conceivable that the passing of the presents in the dynamic conception of time might be predicated upon a becoming which is itself ontologically prior to time.

15 If we were to extrapolate from this argument, and use similar principles by which to explain the becoming, or passing, of time itself, then we might conceive of the event of becoming as consisting in the present's letting go of t_0 and grasping of t_1. It should be noted that neither way of thinking attempts to account for the 'creation' either of t_1, or of the state of affairs that pertain at t_1. Rather, the concept of grasping indicates that t_1, or the state of affairs which pertain at t_1, are already there to be grasped. What is their way of being there? Again, the text offers no illumination. It will only be with the Aristotelian notion of potentiality that an attempt can be made to resolve this question – an attempt that is explicitly criticised subsequently by Bergson (see Keith Ansell Pearson's essay, 'Duration and Evolution', in this collection).

16 In a similar fashion, in the change from t_0 to t_1, there is no time which is *both* t_0 and t_1, nor is there a time which is *neither* t_0 nor t_1; for the first case would dissolve the difference between moments in time, and hence temporal passing in general, whereas the second case would call for the introduction of a further moment of time between t_0 and t_1, which would merely beg the original question. Moreover, the passing from t_0 to t_1 cannot occur in t_0 alone, or in t_1 alone; nor is there a time in which t_0 and t_1 are together. Thus, of time itself, we could equally conclude that 'it is a strange thing in which it is when it is passing'.

17 In his desire to move away from the metaphysical notion of the instant, Aristotle unfortunately moves too close to Zeno in accepting that motion at an instant is impossible; for, as Owen [in 'The Platonism of Aristotle', *Proceedings of the British Academy* 1965, p. 148] and others have argued, this has disastrous effects for any fledgling theory of dynamics. Indeed, it is only with the invention of the differential calculus by Newton and Leibniz, and the ability it offered to derive velocity at an instant, that this legacy has been overcome.

18 As Heidegger writes: 'there is already present a reference to the no-longer and the not-yet. It has dimension within itself; it stretches out toward a not-yet and a no-longer... Because of this *dimensional content* the *now* has within it self the *character of a transition*.' (Martin Heidegger, *The Basic Problems of Phenomenology*, trans. A. Hofstadter (Bloomington: Indiana University Press, 1982), p. 248 [*Die Grundprobleme der Phänomenologie* (Vittorio Klosterman, 1975), pp. 351–2].)

19 *Op. Cit.*, p. 246 [349].

20 *Op. Cit.*, p. 247 [350].

21 *Op. Cit.*, p. 248 [350].

22 For an attempt to develop a thinking of time from the perspective of originary difference, see R. Durie, 'Splitting Time: Bergson's Philosophical Legacy', *Philosophy Today* 44 (2000), pp. 152–168.

23 Citation from Julian Barbour's web site: http//216.92.126.230/ideas.html.

2
The Measure of Time[1]

Henri Poincaré

I

So long as we do not go outside the domain of human consciousness, the notion of time is relatively clear. Not only do we distinguish without difficulty present sensation from the remembrance of past sensations or the anticipation of future sensations, but we know perfectly well what we mean when we say that, of the two conscious phenomena which we remember, one was anterior to the other; or that, of two foreseen conscious phenomena, one will be anterior to the other.

When we say that two conscious facts are simultaneous, we mean that they profoundly interpenetrate, so that analysis cannot separate them without mutilating them.

The order in which we arrange conscious phenomena does not admit of any arbitrariness. It is imposed upon us and we can change nothing about it.

I have only a single observation to add. For an aggregate of sensations to have become a remembrance capable of classification in time, it must have ceased to be actual, we must have lost the sense of its infinite complexity, otherwise it would have remained present. It must, so to speak, have crystallised around a centre of associations of ideas which will be a sort of label. It is only when they thus have lost all life that we can classify our memories in time as a botanist arranges dried flowers in his herbarium.

But these labels can only be finite in the number. On that score, psychological time should be discontinuous. Whence comes the feeling that between any two instants there are others? We arrange our recollections in time, but we know that there remain empty compartments. How could that be, if time were not a pre-existent form in our mind? How could we know there were empty compartments, if these compartments were revealed to us only by their content?

II

But that is not all; into this form we wish to put not only the phenomena of our own consciousness, but those of which other consciousnesses are the theatre. But more, we wish to put there physical facts, these entities with which we people space and which no consciousness sees directly. This is necessary because without it science could not exist. In a word, psychological time is given to us and must serve to furnish scientific and physical time. There the difficulty begins, or rather the difficulties, for there are two.

Think of two consciousnesses, which are like two worlds impenetrable one to the other. By what do we strive to put them into the same mould, to measure them by the same standard? Is it not as if one strove to measure length with a gramme, or weight with a metre? And besides, why do we speak of measuring? We know perhaps that some fact is anterior to some other, but not *by how much* it is anterior.

Therefore two difficulties: (1) Can we transform psychological time, which is qualitative, into a quantitative time? (2) Can we reduce to one and the same measure facts which transpire in different worlds?

III

The first difficulty has long been noticed; it has been the subject of long discussions and one may say the question is settled. *We have not a direct intuition of the equality of two intervals of time*. The persons who believe they possess this intuition are dupes of an illusion. When I say from noon to one the same time passes as from two to three, what meaning does this affirmation have?

The least reflection shows that by itself it has none at all. It will only have that which I choose to give it, by a definition which will certainly possess a certain degree of arbitrariness. Psychologists could have done without this definition; physicists and astronomers could not; let us see how they have managed.

To measure time they use the pendulum and they suppose by definition that all the beats of this pendulum are of equal duration. But this is only a first approximation; the temperature, the resistance of the air, the barometric pressure, make the pace of the pendulum vary. If we could escape these sources of error, we should obtain a much closer approximation, but it would still be only an approximation. New causes, hitherto neglected, electric, magnetic, or others, would introduce minute perturbations.

The Measure of Time

In fact, the best chronometers must be corrected from time to time, and the corrections are made by the aid of astronomical observations; arrangements are made so that the sidereal clock marks the same hour when the same star passes the meridian. In other words, it is the sidereal day, that is, the duration of the rotation of the earth, which is the constant unit of time. It is supposed, by a new definition substituted for that based on the beats of the pendulum, that two complete rotations of the earth have the same duration.

However, the astronomers are still not content with this definition. Many of them think that the tides act as a check on our globe, and that the rotation of the earth is becoming slower and slower. Thus would be explained the apparent acceleration of the motion of the moon, which would seem to be going more rapidly than theory permits, because our watch, which is the earth, is going slow.

IV

All this is unimportant, one will say; doubtless our instruments of measurements are imperfect, but it is sufficient that we can conceive a perfect instrument. This ideal cannot be reached, but it is enough to have conceived it and so put rigour into the definition of the unit of time.

The trouble is that there is no rigour in the definition. When we use the pendulum to measure time, what postulate do we implicitly admit? *It is that the duration of two identical phenomena is the same*; or, if you prefer, that the same causes take the same time to produce the same effects.

And at the first blush, this is a good definition of the equality of two durations. But take care. Is it impossible that experiment may some day contradict our postulate? Let me explain myself. I suppose that at a certain place in the world the phenomenon α happens, causing as consequence at the end of a certain time the effect α'. At another place in the world very far away from the first, the phenomenon β happens, which causes as consequence the effect β'. The phenomena α and β are simultaneous, as are also the effects α' and β'.

Later, the phenomenon α is reproduced under approximately the same conditions as before and *simultaneously* the phenomenon β is also reproduced at a very distant place in the world and almost under the same circumstances. The effects α' and β' also take place. Let us suppose that the effect α' happens perceptibly before the effect β'.

If experience made us witness such a sight, our postulate would be contradicted. For experience would tell us that the first duration $\alpha\alpha'$ is equal to the first duration $\beta\beta'$ and that the second duration $\alpha\alpha'$ is less than the second duration $\beta\beta'$. On the other hand, our postulate would require that the two durations $\alpha\alpha'$ should be equal to each other, as likewise the two durations $\beta\beta'$. The equality and the inequality deduced from experience would be incompatible with the two equalities deduced from the postulate.

Now can we affirm that the hypotheses I have just made are absurd? They are in no way contrary to the principle of contradiction. Doubtless they could not happen without apparently violating the principle of sufficient reason. But to justify a definition so fundamental I should prefer some other guarantee.

V

But that is not all. In physical reality one cause does not produce a given effect, but a multitude of distinct causes contribute to produce it, without our having any means of discriminating the part of each of them.

Physicists seek to make this distinction; but they make it only approximately, and however they progress, they will never make it except approximately. It is approximately true that the motion of the pendulum is due solely to the earth's attraction; but in all rigour every attraction, even of Sirius, acts on the pendulum.

Under these conditions, it is clear that the causes which have produced a certain effect will never be reproduced except approximately. Then we should modify our postulate and our definition. Instead of saying; 'The same causes produce the same effects,' we should say; 'Causes almost identical take almost the same time to produce almost the same effects.'

Our definition, therefore, is no longer anything but approximate. Besides, as M. Calinon very justly remarks in a recent memoir:

> One of the circumstances of any phenomenon is the velocity of the earth's rotation; if this velocity of rotation varies, it constitutes in the reproduction of this phenomenon a circumstance which no longer remains the same. But to suppose this velocity of rotation constant is to suppose that we know how to measure time.[2]

Our definition is therefore not yet satisfactory; it is certainly not that which the astronomers of whom I spoke above implicitly adopt, when they affirm that the terrestrial rotation is slowing down.

What meaning does this affirmation have according to them? We can only understand it by analysing the proofs they give of their proposition. They say first that the friction of the tides producing heat must destroy *vis viva*.³ They invoke therefore the principle of *vis viva*, or of the conservation of energy.

They say next that the secular acceleration of the moon, calculated according to Newton's law, would be less than that deduced from observations unless the correction relative to the slowing down of the terrestrial rotation were made. They invoke therefore Newton's law. In other words, they define duration in the following way: time should be so defined that Newton's law and that of the *vis viva* may be verified. Newton's law is an experimental truth; as such it is only approximate, which shows that we still have only a definition by approximation.

If we now suppose that another way of measuring time is adopted, the experiments on which Newton's law is founded would nonetheless have the same meaning. Only the enunciation of the law would be different, because it would be translated into another language; it would evidently be much less simple. So that the definition implicitly adopted by the astronomers may be summed up thus: Time should be so defined that the equations of mechanics may be as simple as possible. In other words, there is not one way of measuring time more true than another; that which is generally adopted is only more *convenient*. Of two watches, we have no right to say that the one goes true, the other wrong; we can only say that it is advantageous to conform to the indications of the first.

The difficulty which has just occupied us has been, as I have said, often pointed out; among the most recent works in which it is considered, I may mention, besides M. Calinon's little book, the treatise on mechanics of M. Andrade.

VI

The second difficulty has up to the present attracted much less attention; yet it is altogether analogous to the preceding; and even, logically, I should have spoken of it first.

Two psychological phenomena happen in two different consciousnesses; when I say they are simultaneous, what do I mean? When I say that a physical phenomenon, which happens outside of every consciousness, is before or after a psychological phenomenon, what do I mean?

In 1572, Tycho Brahe noticed in the heavens a new star. An immense conflagration had happened in some far distant heavenly body; but it had happened long before; at least two hundred years were necessary for the light from that star to reach our earth. This conflagration therefore happened before the discovery of America. Well, when considering this gigantic phenomenon, which perhaps had no witness, since the satellites of that star were perhaps uninhabited, I say this phenomenon is anterior to the formation of the visual image of the isle of Espanola in the consciousness of Christopher Columbus, what do I mean?

A little reflection is sufficient to understand that all these affirmations have by themselves no meaning. They can have one only as the outcome of a convention.

VII

We should first ask ourselves how one could have the idea of putting into the same frame so many worlds impenetrable to each other. We should like to represent to ourselves the external universe, and only by so doing could we feel that we understood it. We know we never can attain this representation: our weakness is too great. But at least we desire the ability to conceive an infinite intelligence for which this representation would be possible, a sort of great consciousness which should see all, and which should classify all *in its time*, as we classify, *in our time*, the little we see.

This hypothesis is indeed crude and incomplete, because this supreme intelligence would be only a demigod; infinite in one sense, it would be limited in another, since it would have only an imperfect recollection of the past; and it could have no other, since otherwise all recollections would be equally present to it and for it there would be no time. And yet when we speak of time, for all which happens outside of us, do we not unconsciously adopt this hypothesis; do we not put ourselves in the place of this imperfect god; and do not even the atheists put themselves in the place where god would be if he existed?

What I have just said shows us, perhaps, why we have tried to put all physical phenomena into the same frame. But that cannot pass for a definition of simultaneity, since this hypothetical intelligence, even if it existed, would be impenetrable for us. It is therefore necessary to seek something else.

VIII

The ordinary definitions which are proper for psychological time would serve us no better. Two simultaneous psychological facts are so closely bound together that analysis cannot separate without mutilating them. Is

The Measure of Time 31

it the same with two physical facts? Is not my present nearer my past of yesterday than the present of Sirius?

It has also been said that two facts should be regarded as simultaneous when the order of their succession may be inverted at will. It is evident that this definition would not suit two physical facts which happen far from one another, and that, in respect to them, we no longer even understand what this reversibility would be; besides, succession itself must first be defined.

IX

Let us then seek to give an account of what is understood by simultaneity or antecedence, and for this let us analyse some examples.

I write a letter; it is afterwards read by a friend to whom I have addressed it. There are two facts which have had for their theatre two different consciousnesses. In writing this letter I have had the visual image of it, and my friend has had in turn the same visual image in reading the letter. Though these two facts happen in impenetrable worlds, I do not hesitate to regard the first as anterior to the second, because I believe it is its cause.

I hear thunder, and I conclude there has been an electrical discharge; I do not hesitate to consider the physical phenomenon as anterior to the auditory image perceived in my consciousness, because I believe it is its cause.

Behold then the rule we follow, and the only one we can follow: when a phenomenon appears to us as the cause of another, we regard it as anterior. It is therefore by cause that we define time; but most often, when two facts appear to us bound by a constant relation, how do we recognize which is the cause and which the effect? We assume that the anterior fact, the antecedent, is the cause of the other, of the consequent. It is then by time that we define cause. How are we to save ourselves from this *petitio principii*?

We say now *post hoc, ergo propter hoc*; now *propter hoc, ergo post hoc*; shall we escape from this vicious circle?

X

Let us see, not how we succeed in escaping, for we do not completely succeed, but how we try to escape.

I execute a voluntary act A and I feel afterwards a sensation D, which

I regard as a consequence of the act A; on the other hand, for whatever reason, I infer that this consequence is not immediate, but that outside my consciousness two facts B and C, which I have not witnessed, have happened, and in such a way that B is the effect of A, that C is the effect of B, and D of C.

But why? If I think I have reason to regard the four facts A, B, C, D as bound to one another by a causal connection, why range them in the causal order A, B, C, D, and at the same time in the chronological order A, B, C, D, rather than in any other order?

I clearly see that in the act A I have the feeling of having been active, while in undergoing the sensation D, I have that of being passive. This is why I regard A as the initial cause and D as the ultimate effect; this is why I put A at the beginning of the chain and D at the end; but why put B before C rather than C before B?

If this question is posed, the reply ordinarily is: we know that it is B which is the cause of C because we *always* see B happen before C. These two phenomena, when witnessed, happen in a certain order; when analogous phenomena happen without witness, there is no reason to invert this order.

Doubtless, but take care; we never know directly the physical phenomena B and C. What we know are sensations B' and C' produced respectively by B and C. Our consciousness tells us immediately that B' precedes C' and we *suppose* that B and C succeed one another in the same order.

This rule appears in fact very natural, and yet we are often led to depart from it. We hear the sound of the thunder only some seconds after the electrical discharge of the cloud. Of the two flashes of lightning, the one distant, the other near, could not the first be anterior to the second, even though the sound of the second comes to us before that of the first?

XI

Another difficulty; have we really the right to speak of the cause of a phenomenon? If all the parts of the universe are interchained in a certain measure, any one phenomenon will not be the effect of a single cause, but the resultant of several causes infinitely numerous; it is, one often says, the consequence of the state of the universe a moment before. How are we to enunciate rules applicable to circumstances so complex? And yet it is only thus that these rules can be general and rigorous.

The Measure of Time 33

So that we do not lose ourselves in this infinite complexity, let us make a simpler hypothesis. Consider three stars, for example, the sun, Jupiter and Saturn; but, for greater simplicity, regard them as reduced to material points and isolated from the rest of the world. The positions and the velocities of three bodies at a given instant suffice to determine their positions and velocities at the following instant, and consequently at any instant. Their positions at the instant t determine their positions at the instant $t + h$ as well as their positions at the instant $t - h$.

Even more; the position of Jupiter at the instant t, together with that of Saturn at the instant $t + a$, determines the position of Jupiter at any instant and that of Saturn at any instant.

The aggregate of positions occupied by Jupiter at the instant $t + e$ and Saturn at the instant $t + a + e$ is bound to the aggregate of positions occupied by Jupiter at the instant t and Saturn at the instant $t + a$, by laws as precise as that of Newton, though more complicated. Then why not regard one of these aggregates as the cause of the other, which would lead to considering as simultaneous the instant t of Jupiter and the instant $t + a$ of Saturn?

In answer there can only be reasons, very strong, it is true, of convenience and simplicity.

XII

But let us pass to less artificial examples; to understand the definition implicitly supposed by the savants, let us watch them at work and look for the rules by which they investigate simultaneity.

I will take two simple examples, the measurement of the velocity of light and the determination of longitude.

When an astronomer tells me that some stellar phenomenon, which his telescope reveals to him at this moment, happened nevertheless fifty years ago, I seek his meaning, and to that end, I shall ask him how he knows it, that is, how he has measured the velocity of light.

He has begun by *supposing* that light has a constant velocity, and in particular that its velocity is the same in all directions. That is a postulate without which no measurement of this velocity could be attempted. This postulate could never be verified directly by experiment; it might be contradicted by it if the results of different measurements were not concordant. We should think ourselves fortunate that this contradiction has not happened and that the slight discordances which may happen can be readily explained.

The postulate, at all events, resembling the principle of sufficient reason, has been accepted by everybody; what I wish to emphasize is that it furnishes us with a new rule for the investigation of simultaneity, entirely different from that which we have set out above.

This postulate assumed, let us see how the velocity of light has been measured. You know that Roemer used eclipses of the satellites of Jupiter, and researched how much the event fell behind its prediction. But how is this prediction made? It is by the aid of astronomic laws, such as Newton's law.

Could not the observed facts be just as well explained if we attributed to the velocity of light a little different value from that adopted, and supposed Newton's law only approximate? Only this would lead to replacing Newton's law by another more complicated. So for the velocity of light a value is adopted, such that the astronomical laws compatible with this value may be as simple as possible. When navigators or geographers determine a longitude, they have to solve just the problem we are discussing; they must, without being at Paris, calculate Paris time. How do they accomplish it? They carry a chronometer set for Paris. The qualitative problem of simultaneity is made to depend upon the quantitative problem of the measurement of time. I need not take up the difficulties relative to this latter problem, since above I have emphasized them at length.

Or else they observe an astronomical phenomenon, such as the eclipse of the moon, and they suppose that this phenomenon is perceived simultaneously from all points of the earth. That is not altogether true, since the propagation of light is not instantaneous; if absolute exactitude were required, there would be a correction to make according to a complicated rule.

Or else, finally, they use the telegraph. It is clear first that the reception of the signal at Berlin, for instance, is after the sending of this same signal from Paris. This is the rule of cause and effect, analysed above. But how much after? In general, the duration of the transmission is neglected and the two events are regarded as simultaneous. But, to be rigorous, a little correction would still have to be made by a complicated calculation; in practise it is not made, because it would be well within the errors of observation; its theoretic necessity remains undiminished from our point of view, which is that of a rigorous definition.

From this discussion, I wish to emphasize two things: (1) The rules applied are exceedingly various. (2) It is difficult to separate the qualitative problem of simultaneity from the quantitative problem of the measurement of time; no matter whether a chronometer is used, or whether account must be taken of a velocity of transmission, such as that of light, because such a velocity could not be measured without *measuring* a time.

XIII

To conclude: We do not have a direct intuition of simultaneity, nor of the equality of two durations. If we think we have this intuition, this is an illusion. We replace it by the aid of certain rules which we apply almost always without taking account of them.

But what is the nature of these rules? No general rule, no rigorous rule; a multitude of little rules applicable to each particular case.

These rules are not imposed upon us, and we might amuse ourselves in inventing others; but they could not be cast aside without greatly complicating the enunciation of the laws of physics, mechanics and astronomy.

We therefore choose these rules, but not because they are true, but because they are the most convenient, and we may recapitulate them as follows: 'The simultaneity of two events, or the order of their succession, the equality of two durations, are to be so defined that the enumeration of the natural laws may be as simple as possible. In other words, all these rules, all these definitions are only the fruit of an unconscious opportunism.'

Notes

1 Published as Chapter 2 of Henri Poincaré, *La Valeur de la Science* (Paris: Flammarion, 1904); translated by G.B. Halsted as *The Value of Science* (New York: Science Press, 1907; reprinted, New York: Dover Press, 1958).
2 Callinon, *Étude sur les diverses grandeurs* (Paris: Gauthier-Villars, 1897)
3 *Vis viva*: 'the operative force of a moving or acting body, equal to the mass of the body multiplied by the square of its velocity'.

3
Memory of the Present and False Recognition
An Article in the Revue Philosophique, December, 1908[1]

Henri Bergson

The illusion concerning which I am going to submit a few explanatory views is well known. Some one may be attending to what is going on or taking part in a conversation, when suddenly the conviction will come over him that he has already seen what he is now seeing, heard what he is now hearing, uttered the sentence he is uttering – that he has already been here in this very place in which he is now, in the same circumstances, feeling, perceiving, thinking and willing the same things, and, in fact, that he is living again, down to the minutest details, some moments of his past life. The illusion is sometimes so complete that, at every moment while it lasts, he thinks he is on the point of predicting what is going to happen: how should he not know it already, since he feels that he is about to have known it? It is by no means rare for the person under this illusion to perceive the external world under a peculiar aspect, as in a dream; he becomes a stranger to himself, ready to be his double, present as a simple spectator at what he is saying and doing. This 'depersonalisation', to employ a term used to describe the experience by M. Dugas,[2] is not identical with or necessarily a symptom of false recognition; it has, however, a certain relationship to it. Moreover, all the symptoms differ in degree. The illusion, instead of being a complete picture, may often present itself as a mere sketch. But sketch or finished picture, it always bears its original character.

There are on record many descriptions of false recognition. They resemble one another in a striking manner, and are often set forth in identical terms. I have in my possession the self-observation of a literary man, which he specially undertook for me. He was skilled in introspection, had never heard of the illusion of false recognition, and believed himself to be the only person to experience it. His description consists of some dozen sentences, all of which are met with, in almost identical

Memory of the Present and False Recognition 37

words, in published records of other cases. I congratulated myself at first that I had at least obtained a new expression of it, for the author tells me that what dominates the phenomenon is a feeling of 'inevitability', a feeling that no power on earth could stop the words and acts, about to come, from coming. But re-reading the cases recorded by M. Bernard-Leroy,[3] I find in one of them an identical expression: "I was a spectator at my own actions; they were inevitable." Indeed, it is doubtful if there exists another illusion stereotyped with such precision.

I do not include under false recognition certain illusions which resemble it on one side or another, but differ from it in their general aspect. M. Arnaud described in 1896 a remarkable case which he had then had under his observation for three years. Throughout this time the patient had experienced, or *believed he experienced*, continuously the illusion of false recognition, imagining himself living his whole life over again.[4] This case, moreover, is not an isolated one; it seems to approach very nearly a very early case described by Pick,[5] a case described by Kräpelin[6] and also one related by Forel.[7] Reading these cases we are at once aware of something quite different from false recognition. The illusion does not spring up as a sharp and short impression, which surprises by its strangeness. The subject finds, on the contrary, that what he experiences is natural and normal; he sometimes has need of that impression; he seeks it when it fails him, and believes it to be even more continuous than it is in reality. Studying the illusion more closely, we discover other well-marked differences. In false recognition, the illusory memory is never localised in a particular point in the past; it dwells in an indeterminate past – the past in general. In these cases, on the contrary, the patients refer to a particular date the experience they claim already to have had; they are the prey of a real hallucination of memory. They are, it should be observed, all cases of insanity. That of Pick and those of Forel and Arnaud suffer delirious ideas of persecution; that of Kräpelin is a maniac with hallucinations of vision and hearing. Their mental trouble may have some relation to that described by Coriat under the name of 'reduplicative paramnesia',[8] what Pick himself in a more recent work calls 'a new form of paramnesia'.[9] In this last case the subject believed he had already several times lived his actual life. Arnaud's patient had exactly the same illusion.

A more delicate question is raised by the studies of M. Pierre Janet on psychasthenia. In opposition to most authorities, M. Janet considers false recognition a purely pathological state, relatively rare, at any rate vague and indistinct, and holds that it would be unjustifiable on the facts to describe it as a specific illusion of memory.[10] It is in reality concerned, in his view, with a much more general trouble. The 'function of the real' is enfeebled, the patient has not completely succeeded in apprehending the actual; he cannot say with certainty whether now is present, past or even future; he will decide for the past if that idea be suggested in the questions put to him. That psychasthenia, which has been so thoroughly studied by M. Pierre Janet, is the dumping-ground of a host of anomalies, and false recognition is one of them, I do not contest. Nor do I wish to dispute the psychasthenic character of false recognition in all cases. The question is, however, whether the phenomenon, when it is found precise, complete and sharply analysable into perception and memory, when, moreover, it is produced in people who present no other anomaly, has the same internal structure as when it appears with a vague form – rather a tendency or disposition than a definite clean-cut state – in minds which manifest a whole group of psychasthenic symptoms. Suppose that false recognition, considered simply as we know it, a disorder always temporary and never severe, be a means contrived by nature in order to localise at one spot and limit to a few instants and so reduce to its mildest form a certain insufficiency which, were it to spread and, so to speak, be diluted in the whole psychological life, *would be* psychasthenia: we should then expect that this concentration at one spot would give to the resulting state of mind a precision, a complexity and above all an individuality not found generally among patients suffering from general psychasthenia and thereby apt to shape into a vague form of false recognition, as well as into a great many other mental peculiarities, the radical deficiency from which they suffer. The illusion would in such case be a distinct psychological entity, whilst it is not so with general psychasthenic patients. Nothing we are told concerning this illusion in psychasthenic patients need be rejected. But what we have to explain is why there is created the particular feeling of 'already seen' in those cases – numerous, I believe – in which there is the very distinct affirmation of a present perception *and* of a past perception which has been identical with it. It must be borne in mind that many of those who have studied false recognition – Jensen, Kräpelin, Bonatelli, Sander, Anjel and others – were themselves subject to it. They have not limited themselves to collecting cases; as professional psychologists, they have noted what they have themselves experienced. Now all these authorities agree in describing the phenomenon as being clearly a recommencement of the past, a *twofold* phenomenon, which is perception on one side, memory on another, and not a phenomenon of a single aspect, a state in which the

Memory of the Present and False Recognition

reality appears simply in the air, detached from time, perception *or* memory at will. So, without sacrificing anything of what M. Pierre Janet has taught us on the subject of psychasthenia, we have none the less to find a special explanation of the phenomenon distinguished as false recognition.[11]

What is the explanation? In the first place, there is the view of those who hold that false recognition arises from the identification of an actual perception with a former perception really resembling it in its content, or at least in its affective tone. According to some of these authorities (Sander,[12] Höffding,[13] Le Lorrain,[14] Bourdon,[15] Belugou[16]) the past perception belongs to waking experience; according to others (James Sully,[17] Lapie,[18] etc.) to dream experience; according to Grasset,[19] to waking or to dreaming but always to the *unconscious*. According to all, whether they mean the memory of something seen or the memory of something imagined, false recognition is a confused or incomplete recall of a real memory.[20]

This explanation may be accepted within the limits set by several of those who propose it.[21] It applies, in fact, to a phenomenon which resembles false recognition in certain aspects. It has happened to all of us, in the presence of some new scene, to wonder whether we had not seen it before, and on reflection we have found that we had formerly had an analogous perception which presented several features in common with the present experience. But this phenomenon is very different. In false recognition the two experiences appear strictly identical, and we feel indeed that no reflection would reduce the identity to a vague resemblance, because we are not simply beholding the 'already seen'; it is much more than that; we are living through again the 'already lived'. We believe we have to do with the complete reinstatement of one or of several minutes of our past with the totality of their content, presentative, affective, active. Kräpelin, who has insisted on this primary difference, notices still another.[22] The illusion of false recognition comes over a person suddenly and as suddenly vanishes, leaving behind it an impression of dream. We find nothing of the kind in the confusion of a present experience with a former resembling experience – a confusion which is more or less gradual in establishing itself, and more or less easy to dissipate. Let me add (and this is perhaps essential) that such confusion is an error like other errors, a phenomenon localised in the domain of pure intellect. On the contrary, false recognition may disturb our whole personality; it concerns feeling and will as intellect. Whoever experiences it is often the prey of a characteristic emotion, becoming more or less a stranger to himself and, as it were, 'automatised'. In this case, therefore, we have an illusion which includes different elements and which organises them into one single simple effect, a real psychological individuality.[23]

Where must we look for its seat? Is it to be found in an idea, in an emotion or in a state of will?

The tendency to regard it as centred in an idea is characteristic of theories which explain false recognition by bringing in an image supposed to have arisen in the course of perception or a little before it, and to have been at once thrown back into the past. To account for this image, it was first supposed that the brain was double, that it produced two simultaneous perceptions, one of which might in certain cases be lagging by reason of its feebler intensity, and producing the effect of a memory (Wigan,[24] Jensen[25]). Fouillée[26] also speaks of a 'lack of synergy and simultaneity in the cerebral centres', whence is produced a double vision [*diplopie*], 'a pathological phenomenon of echo and internal repetition.' Contemporary psychology is seeking to get away from these anatomical schemes, and the hypothesis of a cerebral duality is now completely abandoned. There remains, then, the theory that the second image may be some part of the perception itself. According to Anjel, we must in fact distinguish two aspects in all perception: the one is the crude impression made on the consciousness, the other the taking possession of that impression by the mind. Ordinarily, the two processes coincide, but if one lag behind the other, a double image results, and this occasions false recognition.[27] Piéron has put forward an analogous idea.[28] Lalande,[29] followed by Arnaud,[30] holds that a scene may produce on us an instantaneous first impression of which we are scarcely conscious, and to this there may succeed a distraction of some seconds, after which the normal perception is established. Should at this moment the first impression come back to us, it would have the effect of a vague memory not localisable in time, and we should then have false recognition. F.W.H. Myers proposed an explanation no less ingenious, founded on the distinction between the conscious and the subliminal ego. The conscious ego receives only a total impression of a scene at which it is present, the details of it being always a little later than those of the external stimulus; the subliminal ego photographs these details one after the other, instantaneously. The latter is therefore in advance of consciousness, and, if suddenly manifested to it, brings a memory of that which the conscious ego is then occupied in perceiving.[31] Lemaître[32] has adopted a position intermediate between those of Lalande and Myers. Before Myers, Dugas

Memory of the Present and False Recognition

had put forward the hypothesis that there is a splitting of the personality.[33] Also, before either of these, Ribot had given great force to the theory of two images by his suggestion that there is in these cases a kind of hallucination more intense than perception and following it: the hallucination throws the perception into the background, so giving it the dim form of a mere remembrance.[34] It is impossible for me to undertake the full examination each of these theories deserves. I am content to say that I accept them in principle. I hold that false recognition implies the very real existence in consciousness of two images, one of which is the reproduction of the other. The great difficulty, in my view, is to explain, first, why one of the two images is thrown back into the past, and, second, why the illusion is continuous. If we take the image thrown back into the past to be anterior to the image localised in the present, if we see in it a first perception less intense, less attended to or less in consciousness than the later perception, we must at least attempt to explain why it takes the form of a memory; but, even then, we have to do only with the memory of a certain moment of the perception; the illusion will not be prolonged and renewed throughout the duration of the perception. If, on the contrary, the two images are formed together, then the continuity of the illusion is easier to understand, but the rejection of one of them into the past calls even more imperatively for explanation. We may indeed ask whether any one of the hypotheses, even of the first kind, really accounts for the throwing back, and whether the feebleness or subconsciousness of a perception suffices to give it the aspect of a memory. In any case, a theory of false recognition must answer at the same time both requirements, and in my view the two requirements must appear irreconcilable so long as the nature of normal memory is not studied from the purely psychological standpoint.

Can we escape the difficulty by denying the duality of the images, by invoking an 'intellectual feeling' of the 'already seen' and supposing it sometimes superadded to our perception of the present, making us believe in a recommencement of the past? This is the idea that has been put forward by M. Bernard-Leroy in an important work.[35] I am quite ready to agree with him that recognition of the present is generally without any calling up of the past. I have myself shown that the 'familiarity' of the objects of daily experience must be ascribed to the automatism of the reactions they provoke, and not to the presence of a memory-image doubling the perception-image. But this feeling of 'familiarity' is surely not what intervenes in false recognition, and M. Bernard-Leroy has himself been at pains to distinguish the one from the other. The feeling of which M. Bernard-Leroy speaks can only be, then, the same as we experience when we say to ourselves, in passing a person on the street, that we must already have met him. But then such feeling is doubtless inseparably bound to a real memory, the memory of that person or of someone else who resembles him: it may be only the vague, almost extinct consciousness of this recollection, together with the nascent and unsuccessful effort to revive it. Then, too, it is significant that in such a case we say, 'I have seen that person somewhere'; we do not say, 'I have seen that person here, in these very circumstances, at a moment of my life indistinguishable from this actually present moment.' If, then, false recognition has its root in a feeling, it is a feeling unique of its kind and it cannot be the feeling of normal recognition wandering over consciousness and deceived as to its destination. Being special, it must depend on special causes, and it behoves us to discover them.

Let us, then, turn to the third group, theories according to which the origin of the phenomenon is to be sought in the sphere of action, rather than in that of feeling or in that of thought. Such is the most recent tendency. Many years ago, I myself called attention to the need of distinguishing various heights of *tension* or *tone* in psychological life. Consciousness, I said, is better balanced the tenser its concentration on action, and more unstable the more it is detended in a kind of dream. Between these two extreme planes – the planc of action and the plane of dream – there are, I added, as many corresponding intermediate planes as there are decreasing degrees of 'attention to life' and adaptation to reality.[36] My suggestions were received with a certain reserve, appearing to some paradoxical. Psychology, however, is now coming nearer and nearer to them, especially since M. Pierre Janet from quite different considerations has reached conclusions altogether in agreement with them. It is in the lowering of such mental tone that, according to the third group of theories, we are to look for the origin of false recognition. In M. Pierre Janet's view, this lowering produces the phenomenon directly by diminishing the effort of synthesis accompanying normal perception, which then takes the aspect of a vague memory or a dream.[37] More precisely, M. Janet thinks that we have to do here with one of the 'feelings of incompleteness' which he has studied in so original a manner. The patient, puzzled at finding that his perception is incompletely real, and therefore incompletely present, hardly knows if he is dealing with the present or the past or even with the future. M. Léon-Kindberg has thought out and developed this idea of diminution of the effort of synthesis.[38] On the other hand, Heymans has tried to show how a 'lowering of psychological energy' might modify the aspect of our habitual environment and communicate the aspect of 'already seen' to events which are happening in it.

> Suppose that our usual surrounding should arouse only very feebly the associations regularly awakened by it. There would then occur precisely what happens when after many years we see again places or objects, hear again melodies, formerly known but long since forgotten... Now, if in such cases of normal recollection we have learnt to interpret the feebler push of associations as a sign of former experiences relating to the same objects as those now present, we may conjecture that in the other cases too, where, following a diminution of psychological energy, the usual surrounding displays a very diminished associative power, we shall have the impression that in it are being repeated, identically, personal events and situations drawn from the depth of a nebulous past.[39]

Lastly, in an elaborate paper written by Dromard and Albès, and in which we find, drawn up as a self-observation, one of the most acute analyses ever given of false recognition,[40] the phenomenon is explained as a diminution of 'attentional tone' which brings about a rupture between the 'lower psychism' and the 'higher psychism'. The lower psychism, functioning without the aid of the higher, perceives the present object automatically, and the higher psychism is then entirely occupied in contemplating the image formed by the lower instead of regarding the object itself.[41]

I may say of these theories, as of the former, that I accept the principle underlying them. It is in a lowering of the general tone of the psychological life that the originating cause of false recognition is to be looked for. The delicate point is to determine the peculiar form which inattention to life takes in this case, and also we must explain why its effect is to mistake the present for a repetition of the past. A mere slackening of the effort of synthesis may indeed give to reality the aspect of a dream – but why should such a dream appear to be the complete repetition of a moment already lived? Even supposing that the 'higher psychism' intervenes in order to superpose its attention on this inattentive perception, all that we should have would be a memory attentively considered, and by no means a perception duplicated with a memory. On the other hand, mere idleness of associative memory, such as Heymans supposes, would simply render difficult the recognition of surroundings: it is a long way from the difficult recognition of something familiar to the memory of a definite past identical in every point with the present. It seems, then, that we must combine two systems of explanation, admit that false recognition is at once a diminution of psychological tension and a duplication of the image, and inquire what must be the diminution which will produce duplication, what the duplication which will simply express diminution. But it would be a mistake to devise any artificial scheme for reconciling the two theories. Let us simply study the mechanism of memory in the two directions indicated, and the two theories will be seen to join together.

However, a remark must first be made concerning all psychological facts which are morbid or abnormal. Among them are some which evidently point to an impoverishment of the normal life. Such are the anaesthesias, the amnesias, the aphasias, the paralyses, all those states, in fact, which are characterised by the loss of particular sensations, particular memories, or particular movements. In order to define these states we simply have to indicate what has disappeared from consciousness. They consist in an absence. We all agree in seeing in them a psychological deficiency.

On the contrary, there are morbid or abnormal states which appear to add something to normal life and enrich it instead of impoverishing it. A delirium, an hallucination, an obsession, are positive facts. They consist in the presence, not in the absence, of something. They seem to introduce into the mind certain new ways of feeling and thinking. To define them, we have to consider what they are and what they bring, instead of what they are not and what they take away. If most of the symptoms of insanity belong to this second category, so also do a great many psychological anomalies and singularities. False recognition is one. As we shall see later, it presents an aspect *sui generis*, far different from that of true recognition.

However, the philosopher may well question whether, in the mental domain, disorder and degeneration can really be capable of creating something, and whether the apparently positive characters which give the abnormal phenomenon an aspect of novelty are not, when we come to study their nature, reducible to an internal void, a shortcoming of normality. Disease, we generally say, is a diminution. True; but this is a vague way of expressing it, and we should indicate precisely, when no actual part of consciousness is missing, wherein consciousness is diminished. I made an attempt of this kind in a former work to which I have already referred. I pointed out that, besides the diminution which affects the *number* of the states of consciousness, there is another which concerns their solidity or *weight*. In the first case, the disorder simply and only eliminates some states without affecting others. In the second, no psychological state disappears but all are affected, all lose something of their ballast, that is to say, of their power of insertion and penetration into the reality.[42] It is the 'attention to life' which is diminished, and the new phenomena which are started are only the visible aspect, the outward appearance of this detachment.

I recognise, however, that even under this form the idea is still too general to be applied to the explanation of particular psychological facts. But it indicates the direction we must follow to find an explanation.

For, if we accept this principle, we shall not, in the case of a morbid or abnormal phenomenon presenting special characters, have to seek any active cause, because the phenomenon, despite appearances, has nothing positive and nothing new about it. It was already being manufactured while the conditions were normal; but it was prevented from emerging, when about to appear, by one of those continually active inhibitory mechanisms which secure *attention to life*. This means that normal psychological life, as I conceive it, is a system of functions, each with its own psychological organ. Were each of these organs to work by itself, there would result a host of useless or untoward effects, liable to disturb the functioning of the others and so upset that adjustable equilibrium by which our adaptation to the environment is continually maintained. But a work of elimination, of correction, of bringing back to the point, is constantly going on, and it is precisely this work which secures a healthy mind. Wherever this work is slackened, symptoms seem to be created, fresh and new, but in reality they were always there, or rather would have been there if nothing had interfered. I quite understand that the investigator should be struck with the *sui generis* character of the morbid facts. As they are complex and yet present a certain order in their complication, his first inclination is to relate them to an acting cause, capable of organising the elements of them. But if, in the mental domain, disease is unable to create, it can only consist in the slackening or stopping of certain mechanisms, which in the normal state prevent others from having their full effect. If this is so, then, in this case *the principal task of psychology is not to explain why certain phenomena are produced in disordered minds, but why they are not found in the normally healthy mind.*

Already I have applied that method to the study of dreams. We are too much inclined to look upon dreams as if they where phantoms superadded to the solid perceptions and conceptions of our waking life, will-o-the-wisps which hover above it. They are supposed to be facts of a special order, to which psychology ought simply to devote a special chapter and then be quit of them. And it is natural they should appear so, because the waking state is what maters to us, whilst the dreaming state is most foreign to action and most useless. From the practical point of view we come to regard it as an accident. But let us set aside this preconceived idea, and the dream-state will then be seen, on the contrary, to be the substratum of our normal state. The dream is not something fantastic hovering above and additional to the reality of being awake; on the contrary, that reality of the waking state is gained by limitation, by concentration and by tension of a diffuse psychological life, which is the dream-like. In a sense, the perception and memory we exercise in the dream-state are more natural than those in the waking state: there does consciousness disport itself, perceiving just to perceive, remembering just to remember, with no care for life, that is, for the action to be accomplished. But the waking state consists in eliminating, in choosing, in concentrating unceasingly the totality of the diffuse dream-life at the point where a practical problem is presented. To be awake means to will. Cease to will, detach yourself from life, disinterest yourself, and by that mere abstention you pass from the awake-self to the dream-self – less *tense* but more *extended*. The mechanism of the awake-state is, then, the more complex, more delicate and more *positive* of the two, and it is the awake-state, rather than the dream-state, which requires explanation.

Now, if dreams are in every respect an imitation or counterfeit of insanity, we may expect our remarks on dreams to apply as well to many forms of insanity. Of course, we must avoid approaching the study of mental diseases with anything like a stereotyped system. It is doubtful if all the phenomena of insanity are to be explained on one and the same principle. And for many of them, still undefined, it is hardly possible yet to attempt an explanation. As I said at first, I offer my view simply as a methodological indication, with no other object than that of pointing a direction for theoretical inquiry. There are, however, some pathological or abnormal facts to which I believe it is now applicable. One of the chief of these is false recognition. For the mechanism of perception and the mechanism of memory seem to me such that false recognition would arise naturally from the joint play of the two faculties, were there not a special mechanism intervening at the same time in order to prevent it. The important thing to know, then, is not why it arises in certain persons at particular moments, but why it is not being produced at every moment in everybody.

How is a memory [*le souvenir*] formed?[43] Let it first be clear, however, that the memories of which I am going to speak are always psychological, although they may be more often unconscious or semi-conscious. Concerning memories considered as traces left in the brain, I have given my view in *Matter and Memory*, the work to which I have had frequent occasion to refer.⁴⁴ I have attempted there to prove that the various memories [*mémoires*] are indeed localised in the brain, in the meaning that the brain possesses for each category of memory-images a special contrivance whose purpose is to convert the pure memory into a nascent perception or image; but if we go further than this, and suppose every memory to be localised in the matter of the brain, we are simply translating undoubted psychological facts into very questionable anatomical language, and we end in consequences which are contradicted by observation. Indeed, when we speak of our memories, we think of something our consciousness possesses or can always recover by drawing in, so to speak, the thread which holds it. The memory, in fact, passes to and fro from consciousness to unconsciousness, and the transition from one to the other is so continuous, the limit between the two states so little marked, that we have no right to suppose a radical difference of nature between them. It is then memory in this sense which is going to occupy us. On the other hand, let us agree to call 'perception' the consciousness of anything that is present, whether it be an internal or external object. Both definitions being granted, I hold that *the formation of memory is never posterior to the formation of perception; it is contemporaneous with it.* Step by step, as perception is created, it is profiled in memory, which is beside it like a shadow is next to a body. But, in the normal condition, there is no consciousness of it, just as we should be unconscious of our shadow were our eyes to throw light on it each time they turned in that direction.

For suppose that memory is not created along with perception: at what moment will it begin to exist? In order to arise, does it wait until the perception has vanished? This is what we usually suppose, whether we think unconscious memories are psychological states or cerebral modifications. In the one case we suppose a present psychological state, the perception, then, when that no longer exists, the memory of that absent perception. In the other case, we think that when certain cells come into play there is perception, and that the action of those cells has left traces so that, when the perception has vanished, there is memory. But, if things happen in this way, the course of our conscious existence must be composed of clear-cut states, each of which must begin objectively, and also objectively end. Now, is it not clear that dividing psychological life into states, as we divide a play into scenes, is relative to the varied and changing interpretations we give of our past and has nothing absolute about it? According to the point of view in which I am placed, or the centre of interest which I choose, I divide yesterday differently, discovering several very different series of situations or states of it. Though these divisions are not all equally artificial, not one existed in itself, because the unrolling [*déroulement*] of psychological life is continuous. The afternoon I happen to have spent in the country with friends has broken up into luncheon + walk + dinner, or into conversation + conversation + conversation, etc., and of none of these conversations, each one encroaching upon another, could it be said that it forms a distinct entity. Scores of systems of disconnecting are possible; no system corresponds with the plain connections of reality. What right have we, then, to suppose that memory [*la mémoire*] chooses one particular system, or that it divides psychological life into definite periods and awaits the end of each period in order to rule up its accounts with perception?

Is it alleged that the perception of an external object begins when the object appears, and ends when it disappears, and that therefore we can, in this case at least, mark the precise moment when memory replaces perception? But this is to ignore the fact that the perception is ordinarily composed of successive parts, and that these parts have just as much individuality, or rather just as little, as the whole. Of each of them we can as well say that its object is disappearing all along: how, then, could the memory arise only when everything is over? And how could memory [*la mémoire*] know, at any particular moment of the operation, that everything was not over yet, that perception was still incomplete?

The more we reflect, the more impossible it is to imagine any way in which the memory can come into being if it is not created step by step with the perception itself. Either the present leaves no trace in memory [*la mémoire*], or it splits at every instant [*se dedouble à tout instant*], its very upsurge [*jaillissement*] being in two jets, symmetrical, one of which falls back [*retombe*] towards the past whilst the other springs forward [*s'élance*] towards the future. But the forward-springing one, which we call perception, is that alone which interests us. We have no need of the memory of things whilst we hold the things themselves. Practical consciousness throwing this memory aside as useless, theoretical reflection holds it to be non-existent. Thus the illusion is born that memory *succeeds* perception.

But this illusion has another source, deeper still.

The main cause is that the reanimated and conscious memory produces on us the effect of being the perception itself and nothing other than the perception, resuscitated in a more modest form. Between perception and memory there seems to be a difference of intensity or degree, but not of nature. The perception being defined a strong state and the memory a weak state, the memory of a perception then necessarily being nothing else than that same perception weakened, it seems

Memory of the Present and False Recognition

to us that memory [*la mémoire*], in order to register a perception in the unconscious, must wait for the perception to put itself to sleep in memory. And so we suppose the memory of a perception can neither be created with this perception nor developed at the same time as it.

But the theory that present perception is a strong state, and revived memory a weak state, that perception passes into memory by way of diminution, is contradicted by the most elementary observation of fact. Take an intense sensation and make it gradually decrease to zero. If there is only a difference of degree between the memory of the sensation and the sensation itself, the sensation will become memory before it disappears. Now, a moment may come when you are unable to say whether you are dealing with a weak sensation felt or a weak sensation imagined, but the weak state never becomes the memory, thrown back into the past, of the strong state. The memory is, then, a different thing.

The memory of a sensation is capable of *suggesting* the sensation, I mean of causing it to be born again, weak at first, then stronger and stronger in proportion as the attention is more fixed upon it. But the memory is distinct from the sensation it suggests; and it is precisely because we feel it behind the sensation it suggests, as the hypnotiser is behind the hallucination he provokes, that we localise in the past the cause of that which we feel. Sensation is essentially what is actual and present; but the memory which suggests it from the depths of the unconscious, hardly emerging upward, presents itself with that *sui generis* power of suggestion which is the mark of that which is no more and would fain exist again. Hardly has the suggestion touched the imagination than the thing suggested is outlined in its nascent state, and this is why it is so difficult to distinguish between a weak sensation felt and a weak sensation which we remember without dating it. But the suggestion is in no degree what it suggests. The pure memory of a sensation or of a perception is not a degree of the sensation or the perception itself. To suppose it so would be like saying that the word of the hypnotiser, in order to suggest to the hypnotised patient that he has in his mouth sugar or salt, must already itself be a little sugared or salted.

If we try to discover the source and purpose of this illusion, we find that innate in our mind is the need to represent our whole inner life as modelled on that very small part of oneself which is inserted into the present reality, the part which perceives it and acts upon it. Our perceptions and our sensations are at once what are clearest in us and most important for us; they note at each instant the changing relation of our body to other bodies; they determine or orient our conduct. Thence our tendency to see in the other psychological facts nothing but perceptions or sensations obscured or diminished. Those, indeed, among us who resist this tendency, who believe thought to be something other than a play of images, nevertheless have some trouble in persuading themselves that the memory of a perception is radically different form the perception itself. The memory must, at any rate, it seems to them, be expressible in terms of perception, must be obtained by some operation effected upon the image. What is the operation? We can say *a priori* that the operation must effect an alteration in the *quality* of the content of the image, or in its *quantity*, or in both at once. Now, it is certainly not in the quality, since memory must represent the past to us without altering it. It must be then in the quantity. But quantity, in its turn, may be extensive or intensive, for the image comprehends a definite number of parts and it presents a certain degree of force. Consider the first alternative. Does memory modify the extension of the image? Evidently not, for if it added anything to the past, it would be unfaithful to it, and if it subtracted something from the past, it would be incomplete. What is left therefore is that the modification bears on the intensity; and as it is evidently not an increase, it must be a diminution. Such is the instinctive, scarcely conscious dialectic by which we are led, from elimination to elimination, to see in the memory an enfeeblement of the image.

When once we have reached this conclusion, our whole psychology of memory is inspired by it; even our physiology feels the effect of it. In whatever way we then represent the cerebral mechanism of perception, we see in memory nothing but the same mechanism set going anew, an attenuated repetition of the same fact. For all that, experience appears to say the opposite. Experience shows us that it is possible to lose visual memories without ceasing to see and auditory memories without ceasing to hear, that psychological blindness and deafness do not necessarily imply loss of sight or of hearing: how would this be possible if perception and memory were concerned with the same centres, and put in play by the same mechanisms? But we turn aside or pass on rather than assent to a radical distinction between perception and memory.

In so far, then, as our reason reconstructs psychological life out of conscious states sharply delineated, and in so far as it judges that all these sates are expressible in terms of images, it is following two paths which converge in making memory an enfeebled perception, something which follows perception instead of being contemporaneous with it. Set aside this natural dialectic of the intellect, convenient though it be for expression in language, possibly indispensable in practice, but not suggested by

Memory of the Present and False Recognition 51

inward observation: memory will appear as doubling perception at every instant, arising with it, developing itself at the same time, and surviving it precisely because it is of a quite different nature.

What then is memory? Every clear description of a psychological state is made up of images, and we are saying that the memory of an image is not an image. The pure memory, then, can only be described in a vague manner and in metaphorical terms. Let me repeat, then, an explanation I suggested in *Matter and Memory*,[45] which is that the memory is to the perception as the image reflected in the mirror is to the object in front of it. The object can be touched as well as seen; it acts on us as well as we on it; it is pregnant with possible actions, it is *actual*. The image is *virtual*, and though it resembles the object, it is incapable of doing what the object does. Our actual existence, then, while it is unrolled [*déroule*] in time, doubles itself in this way with a virtual existence, a mirror image. Each moment of our life offers two aspects: it is actual and virtual; perception on one side and memory on the other. It splits [*Il se scinde*] as and when it is posited. Or rather, it consists in this very splitting, for the present instant, always running on [*en marche*], fleeting limit between the immediate past which is no more and the immediate future which is not yet, would be reduced to a simple abstraction if it were not precisely the moving mirror which unceasingly reflects perception in memory.

Let us imagine a mind which becomes conscious of this doubling. Suppose the reflection of our perception and of our action comes to consciousness not when the perception is complete and the action accomplished, but in step with our perceiving and acting. We will then see, at one and the same time, our real existence and its virtual image, the object on one side and the reflection on the other. Moreover, the reflection cannot be confused with the object, for the former has all the characters of perception whilst the latter is already memory: if it were not already memory, then it could never become so. Later on, when performing its normal function, it will represent our past to us with the mark of the past; discerned at the moment when it is formed, it already appears to us with the mark of the past, which is constitutive of its essence. What is this past? It has no date and will have none; it is the past *in general*, it cannot be any past in particular. At a pinch, if it merely consisted in a specific spectacle perceived, in a certain emotion experienced, one could be deceived, and believe that one had previously perceived what one is perceiving, experienced what one is experiencing. But it is something quite different. What is doubled at each instant in perception and memory is the totality of what we are seeing, hearing and experiencing, all that we are with and all that surrounds us. If we become conscious of this doubling, it is the entirety of our present which must appear to us at once as perception and as memory. And yet we know full well that no life goes twice through the same moment of its history, and that time does not remount [*remonte*] its course. What is to be done? The situation is strange, paradoxical. It disrupts everything to which we have become habituated. A memory is there: it is indeed a memory, since it bears the characteristic mark of states which we commonly call by this name, and which are only delineated for consciousness when their object has disappeared. And yet it does not present to us something which has been, but simply something which is; it advances *pari passu* with the perception which it reproduces. It is, in the actual moment, a memory of that moment. It is of the past in its form and of the present in its matter. It is a *memory of the present*.

Step by step, as the situation progresses, the memory, which keeps pace with it, gives to each of its stages the aspect of 'already seen' ['*déjà vu*'], the feeling of already known [*déjà connu*]. But the situation, even before it has come to an end, seems to us something which must form a whole, being cut out of the continuity of our experience by the interest of the moment. Now, how could we already have lived a part of the situation if we had not lived the whole of it? Could we recognise what is being unrolled if we did not know what is still rolled up? Are we not able at each moment to anticipate at least the following moment? The instant which is about to come is already broken into by the present instant; the content of the first is inseparable from the content of the second; if, as is beyond doubt, the present instant is a recommencement of my past, then is not the instant to come equally a recommencement of my past? If I recognise the present instant, am I not as surely going to recognise the instant to come? In this way I find myself unceasingly in the attitude, towards what is on the point of happening, of a person who will recognise, and who consequently knows. But this is only the *attitude* of knowledge; it is the form without the matter. As I cannot predict what is going to happen, I see very well that I do not know it; but I foresee that I am going to have known it, in the sense that I will recognise it when I perceive it; and this recognition to come, which I feel to be inevitable in virtue of the impetus[46] taken up all along by my faculty of recognising, exercises in advance a retroactive effect on my present, placing me in the strange situation of a person who feels he knows that of which he knows he is ignorant.

Suppose we catch ourselves repeating mechanically something we once knew by heart but had long forgotten. As we recognise each word the moment we pronounce it, we have a feeling that we possess it before pronouncing it; and yet we only retrieve it when we pronounce it. Whoever becomes conscious of the continuous doubling of their present in perception and in memory will be in the same state. If even slightly capable of self-analysis, she will compare herself to an actor playing her part automatically, both listening to herself and regarding herself play. The more deeply she analyses her experience, the more she will split into two personages, the one of which is thus given as a spectacle to the other. On the one hand, she knows that she continues to be what she was, a self who thinks and who acts in conformity with what the situation demands, a self inserted in real life and adapting itself to it by a free effort of will; of this she is assured by her perception of the present. But the memory of this present, which is equally there, makes her believe that she is repeating entirely what has already been said, seeing exactly what has been already seen, and in this way transforming herself into an actor who recites a role. Of these two different selves, one, conscious of its liberty, erects itself into an independent spectator of a scene which the other plays out in a mechanical manner. But this doubling does not go through to the end. It is rather an oscillation between two standpoints from which the person views themselves, a coming and going of the mind between perception which is nothing other than perception, and perception which is doubled with its own memory. The first involves the habitual feeling we have of our freedom and inserts itself quite naturally into the real world; the second makes us believe we are repeating a part we have learned, converts us into automata, transports us into a theatrical world or a dream-world. Whoever has experienced during a few instants a pressing danger, from which they have only been able to escape by a rapid series of actions imperatively called for and boldly executed, knows something of the kind. It is a doubling rather virtual than real. We act and yet 'are acted'. We feel that we choose and that we will, but that we are choosing what is imposed upon us and willing the inevitable. Because of such a compenetration of states which melt into one another, and which even coincide in immediate consciousness, but which remain nevertheless logically incompatible, reflective consciousness will represent to itself a doubling of the self into two different personages, one of which appropriates all that has to do with freedom, while the other keeps to necessity – the one, a free spectator, regards the other automatically playing their role.

We have described the three principle aspects of what would appear to us, in our normal state, if we were able to witness the splitting of our present. But these are precisely the characteristics of false recognition. We find them more accentuated the more definite the phenomenon is, the more complete it is, the more profoundly analysed it is by the person who experiences it.

First, many have indeed spoken of a feeling of automatism, and of a state comparable to that of an actor playing a role. What is said and what is done, what the person him or herself says and does, appear 'inevitable'. They are witnessing their own movements, thoughts and actions.[47] Things happen *as if* he or she were doubled, without, however, their being really doubled. One of these subjects writes: 'This feeling of doubling only exists in the sensation; the two persons are only one from the material standpoint.' He means, to be sure, that he experiences a feeling of duality, but accompanied by the awareness that there is but one and the same person.[48]

Second, as I said at the outset, the subject often finds himself in the singular state of mind of a person who believes he knows what is going to happen, while feeling himself unable to predict it: 'It always seems to me,' said one, 'that I am foreseeing what is going to happen, yet I cannot really announce it.' Another recalls what is going to happen, 'as one recalls a name which is at the uttermost limits of memory.'[49] One of the earliest observations is that of one who believed he knew beforehand what the people around him would do.[50] We find in this another characteristic of false recognition.

But, third, the most general characteristic of all is the one to which we first drew attention: the memory evoked is a memory hanging in the air, with no point of attachment in the past. It does not correspond with any previous experience. One knows this, is convinced of it, and this conviction is not the effect of reason: it is immediate. It merges with the feeling that the memory evoked must simply be a duplicate of the actual perception. Is it, then, a 'memory of the present'? If one does not use these words, it is without doubt because the expression appears contradictory, because one does not conceive of memory otherwise than as a repetition of the past, because one cannot admit that a representation might bear the mark of the past independently of that which it represents, and lastly, because one is a theoretician without knowing it, one maintains that every memory is posterior to the perception which it reproduces. But one comes close to saying such a thing; one speaks of a past which is not separated from the present by any interval: 'I felt within me a sort of release [*déclenchement*] which removed all the past lying between that minute of long ago and the minute in which I then was.'[51] The phenomenon is

Memory of the Present and False Recognition

indeed well characterised in this way. When one speaks of 'false recognition', one must specify that it is a matter of a process which does not really counterfeit true recognition and which does not give an illusion. What, indeed, is normal recognition? It may come about in two ways, either through a feeling of familiarity which accompanies the present perception, or through the evoking of a past perception which the present perception seems to repeat. But false recognition is neither one nor other of these two operations. What characterises the first kind of recognition is that it excludes any recall of a determinate personal situation in which the recognised object had formerly been perceived. My desk, my table, my books form around me an atmosphere of familiarity only so long as they do not call up the memory of a determinate event of my history. If they evoke the precise memory of an incident in which they have been mixed up, I recognise them as being part of that incident, but this recognition is superadded to the first and is fundamentally distinct from it, as distinct as the personal is from the impersonal. But false recognition is something other than this feeling of familiarity. It always bears on a personal situation, which one is convinced reproduces another personal situation, just as precise and just as determinate. That would leave recognition of the second kind, that which implies the recall of a familiar situation similar to the one actually experienced. But it should be noted that it is always a matter, in such cases, of similar situations and not of identical situations. Recognition of the second kind is brought about as much by the representation of that which differentiates the two situations at the same time as it is by what is common to them. If I attend for a second time a play, I recognise one by one each of the words, each of the scenes; I recognise at last the whole of the piece, and I recall having seen it already; but then I had a different seat, I had other neighbours, I was taken up with other preoccupations; in any case, I could not have been then what I am today, since I have lived in the interval. If then the two images are the same, they are not presented in the same frame, and the vague feeling of the difference of frames surrounds, like a fringe, the consciousness I have of the identity of the images and allows me at each instant to distinguish them. As opposed to this, in false recognition, the frames are identical, as the images themselves are. I witness the same spectacle with the same sensations, the same preoccupations: in short, I am at this moment in the same spot, on the same date, in the same instant of my history where I was then. It is then hardly fit to speak here of illusion, since illusory knowledge is the imitation of real knowledge, and since the phenomenon with which we are dealing imitates no other phenomenon of our experience. And it is hardly fit to speak of false recognition, since there is no true recognition, of the one kind or the other, of which it would be the exact counterfeit. In reality, it is a matter of a phenomenon unique of its kind, the very phenomenon that would be produced by the 'memory of the present' were it to arise suddenly from the unconscious where it must stay. It might give the impression of a memory, since memory bears a distinctive mark, other than that of perception; but it could not be brought back to any past experience, because each of us knows very well that we do not live twice through the same moment of our history.

It remains for me to say why this memory stays ordinarily concealed, and how it is revealed in extraordinary cases. In a general manner, *as of right*, the past only returns to consciousness in the measure to which it can aid the comprehension of the present and the anticipation of the future: it is a precursor [*éclaireur*] of action. One goes wrong when one studies the functions of representation in an isolated state, as if they were an end in themselves, as if we were pure minds occupied in viewing the passage of ideas and images. The present perception would then attract to itself a similar memory without any forethought for utility, for nothing, for the pleasure – for the pleasure of introducing into the mental world a law of attraction analogous to that which governs the world of bodies. We do not, of course, contest the 'law of similarity', but, as we have remarked elsewhere,[52] any two ideas and any two images taken at random, however removed one might suppose, will always resemble one another in some way, since one can always find a common genus into which they can be made to fit: so that any perception could recall any memory, if there were nothing more than a mechanical attraction of like by like. The truth is that if a perception recalls a memory, it is in order that the circumstances which have preceded, accompanied and followed the past situation should throw some light on the actual situation, and show how to negotiate it. Thousands and thousands of memories may be evoked by resemblance, but the memory that tends to reappear is the one which resembles the perception in a certain particular way, the one which may illuminate and direct the action in preparation. And this memory need not, to be sure, show itself: it suffices that it recalls, without showing itself, the circumstances which have been given in contiguity with it, those which preceded and those which followed it, in short, what it is important to know in order to comprehend the present and to anticipate the future. One might even conceive that nothing at all need be manifested to consciousness, and that the conclusion alone should appear, that is to say, the precise suggestion of a certain approach to take. It is probably in this way that things proceed in the majority of animals. But the more consciousness develops, the more it illuminates the operation of memory [*la mémoire*] and the more too it lets association by resemblance, which is its means, shine through from behind association by contiguity, which is its end. Once this is settled in consciousness, it allows the introduction, by virtue of their resemblance, of a crowd of superfluous memories, lacking in actual interest: this explains how we are able to dream a little while active; but it is the requirements of action which determine the laws of recollection; they alone hold the keys of consciousness, and the dream-memories only gain access through taking advantage of what is lax, what is ill-defined, in the relation of resemblance which grants permission for entry. In short, if the totality of our memories exerts at each instant a push from the depths of the unconscious, consciousness, attentive to life, only legally admits those which are able to combine with the present action, although a lot of others insinuate themselves under cover of this general condition of resemblance which is laid down as a requisite for admittance.

But what could be more useless for present action than memory of the present? Any other memory could be invoked with more right, since it at least brings with it some information, even though it might be without actual interest. We learn nothing from memory of the present alone, since it is nothing other than the double of perception. We have the real object: what are we to do with its virtual image? We may as well release the quarry for the shadow.

This is why there is no memory from which our attention is turned away more obstinately.

The attention at issue here is not that individual attention whose intensity, direction and duration changes from person to person. It is rather, so to speak, generic attention, an attention naturally turned towards certain regions of psychological life and naturally turned away from others. Within each of these regions, our individual attention may, undoubtedly, be directed by its own whim, but then it will simply come to superimpose itself upon the former attention, as the choice that the individual eye makes of such and such an object for viewing is superimposed upon the choice which the fully developed human eye has made, once and for all, of a certain determinate region of the spectrum in which it sees light. But, if a slight weakening of individual attention is nothing other than a normal distraction, the total failure of generic attention finds its expression as a pathological or abnormal fact.

False recognition is such an anomaly. It is due to a temporary enfeebling of general attention to life: the look of consciousness, no longer maintaining its natural direction, allows itself to be distracted in consideration of what is without interest for perception. But what is to be understood here by 'attention to life'? What is the special type of distraction which ends in false recognition? Attention and distraction are vague terms: can they be defined more precisely in this particular case? We shall attempt to do so, without claiming, however, to attain in so obscure a subject complete clarity and definite precision.

We do not notice enough that our present is, above all, an anticipation of our future. The vision which reflective consciousness gives us of our interior life is, undoubtedly, that of one state succeeding another state, each of these states commencing at one point, finishing at another, and provisionally self-sufficient. In this way, the view of reflection prepares the view of language; it distinguishes, separates and juxtaposes; it is only at its ease in the definite and in the immobile; it stops at a static conception of reality. But immediate consciousness grasps quite another thing. Immanent in the interior life, it feels it rather than sees it; but it feels it as a movement, as a continuous encroachment on a future which recoils incessantly. This feeling becomes much clearer when it is a matter of an activity which we are determined to complete. The end of the operation appears to us immediately, and, throughout the whole time during which we are acting, we are conscious not so much of our successive states as of the decreasing distance between our actual position and the end towards which we are approaching. This end, moreover, does not appear otherwise than as a provisional end; we know that there is another thing behind; in the spring which we take to clear the first obstacle we already prepare to leap a second, in expectation of others which will follow indefinitely. Again, when we listen to a sentence, we do not need to pay attention to each word taken in isolation: it is the sense of the whole which matters to us; from the beginning, we reconstruct this sense hypothetically; we cast our mind in a certain general direction, and are left to inflect this direction in various ways, to the extent that the sentence, as it unfolds, pushes our attention towards one sense or another. Here again, the present is perceived in the future upon which it encroaches, rather than being grasped in itself. This impetus gives to all the psychological states, which it makes us pass or step over, a particular aspect, to the constancy of which we are so accustomed that we perceive its absence, when it is lacking, rather than its presence. Each of us may have noticed the strange character which a familiar word sometimes takes when we fix our attention on it. The word then appears as new, and indeed it is; for until then, our consciousness had never made it a stopping point; it was passed over in order to arrive at the end of the sentence. It is not so easy to constrain the impetus of our psychological life as completely as that of our speech; but, whenever the general impetus is weakened, the situation passed through must consequently appear as bizarre as the sound of a word which is immobilised in the course of the movement of a sentence. It is no longer part and parcel of real life. Looking through our past experiences for what resembles it most, it is with a dream that we shall compare it.

Now, it is remarkable that most of the subjects, describing what is felt during and after false recognition, speak of an impression of dreaming. M. Paul Bourget says that the illusion is accompanied by 'a kind of unanalysable feeling that reality is a dream'.[53] And an English writer some years ago, describing his own experience, applied the epithet 'shadowy' to the whole phenomenon, adding that it appeared later, when it was recollected, as 'the half-forgotten relic of a dream'. Thus we have observers, unknown to one another, speaking different languages, expressing themselves in equivalent terms. The impression of dream is, then, almost general.

But it should also be remarked that those persons who undergo false recognition are often liable to find a familiar word strange. An inquiry instituted by G. Heymans has shown that these two dispositions are connected to one another.[54] The author adds, with good reason, that current theories of the first phenomenon do not explain why it is associated with the second.

In these conditions, should we not search for the initial cause of false recognition in a momentary halt in the impetus of our consciousness, a halt that, to be sure, changes nothing in the materiality of our present, but that detaches it from the future of which it is an integral part and from the action which would be its normal conclusion, so giving it the aspect of a mere picture, of a spectacle which one offers to oneself, of a reality transposed into a dream? Allow me to describe a personal impression. I am not subject to false recognition, but I have tried very often, since I have been studying it, to place myself in such a state of mind as described by observers, and to induce in myself the phenomenon experimentally. I have never quite succeeded; but I have obtained, on various occasions, something approaching it, but which is quite fugitive. I should find myself in the presence of a scene which is not only new to me but which contrasts sharply with the course of my habitual life. It might be, for example, a scene where I am on a journey, especially if the journey is improvised. The first condition is, then, that I experience a certain quite specific astonishment, which I shall call the *astonishment at finding myself there*. On to this astonishment there comes to be grafted a rather different feeling, yet which is related to it: the feeling that the *future is closed*, that the situation is detached from everything although I am attached to it. In the degree to which these two emotions interpenetrate, the reality loses its solidity and my perception of the present tends to double itself with some other thing, which is behind it. Is this the 'memory of the present' showing through? I do not venture to say so; but it does indeed seem as if I am on the way to false recognition, and that with a little more I would arrive at it.

Now, why does *memory of the present* wait, before revealing itself, for the *impetus of consciousness* to weaken or halt? We know nothing of the mechanism by which a representation comes up from the unconscious or falls back into it. All that we can do is have recourse to a provisional schema by which to symbolise the operation. Let us return to one that we used before. We represent the totality of unconscious memories as pressing against consciousness – consciousness which only allows to pass those which can combine for action. The memory of the present makes an effort, as do the others; it is, moreover, closer to us than the others; leaning over our perception of the present, it is always on the point of entering. Perception only escapes by a continual movement forward, which maintains the separation. In other words, a memory is never actualised other than by the intermediary of a perception: the memory of the present would thus penetrate consciousness if it were able to insinuate itself into the perception of the present. But this latter is always in advance of it: thanks to the impetus which animates it, perception is less in the present than in the future. Suppose that, all of a sudden, the impetus is halted: the memory catches up with the perception, the present is recognised at the same time as it is cognised.

False recognition, then, would indeed be the most inoffensive form of inattention to life. A constant lowering of the tone of fundamental attention is expressed by psychological disorders more or less profound and enduring. But it may happen that this attention is maintained ordinarily at its normal tone, and that its insufficiency manifests itself in a quite different manner, by these halts in functioning, generally very short, spaced out at long intervals. When the halt occurs, false recognition comes to consciousness, covers it over for some instants then falls back immediately, like a wave.

We conclude with a final hypothesis, at which we hinted at the beginning of our work. If inattention to life is able to take two forms which are not of equal seriousness, should one not in principle suppose that the second, the more benign, is a means of preserving oneself from the other? Where an insufficiency of attention would by its expression risk a definitive passage from a state of wakefulness to a state of dreaming, consciousness localises the evil at a few points where it arranges for attention so many short halts; attention is thus able to maintain itself, for the rest of the time, in contact with reality. Certain very precise cases of false recognition confirm this hypothesis. The subject begins by feeling detached from everything, as in a dream: false recognition occurs immediately afterwards, when the subject begins to regain their self-control.[55]

Such then seems to be the disorder of will which occasions false recognition. It seems, at least, to be its initial cause. As for its direct cause, it must be sought elsewhere, in the combinatory play of perception and memory [*la mémoire*]. False recognition results from the natural functioning of these two faculties given up to their own forces. It would take place in each instant if the will, unceasingly tending towards action, did not prevent the present from returning back on itself through its indefinite pushing on into the future. The *impetus of consciousness*, which manifests the impetus of life, escapes analysis by its simplicity. We can however study, in the moments when it relents, the conditions of mobile equilibrium which it had till then maintained, and in this way analyse a manifestation which lets its essence show through.

Notes

1 Subsequently collected in Bergson's *L'Énergie spirituelle* (Paris, 1919), translated as *Mind-Energy* by H. Wildon Carr (London: Macmillan, 1920). The essay as it appears in this volume is a revised version of Carr's translation. [Editor's note]
2 'Un Cas de dépersonnalisation', *Revue Philosophique* (1898), pp. 500–507.
3 *L'Illusion de fausse reconnaissance* (Paris 1898), p. 176.
4 *Annales medico-psychologiques* (1896), pp. 455–470.
5 *Archiv für Psychiatrie* (1876), pp. 568–574.
6 *Ibid.* (1887), p. 428.
7 *Das Gedächtnis und seine Abnormitäten* (Zürich, 1885), pp. 44–45.
8 *Journal of Nervous and Mental Diseases* (1904), pp. 577–78, 639–659.
9 *Jahrbücher für Psychiatrie und Neurologie* (1901), pp 1–35.
10 *Les Obsessions et la psychasthénie* (1903), vol. I. p. 287 ff; cf. *Journal de psy-*

chologie (1905), pp. 139–166.
11 We may note that most authorities regard false recognition as a very widespread illusion. Wigan thought everyone subject to it. Kräpelin calls it a normal phenomenon. Jensen declares that almost any one, attentive to themselves, may experience the illusion.
12 *Archiv für Psychiatrie* (1874), pp. 244–253.
13 *Psychologie*, pp 166–167
14 *Revue Philosophique* (1894), pp. 208–210.
15 *Revue Philosophique* (1893), pp 629–631.
16 *Revue Philosophique* (1907), pp 282–284.
17 *Les Illusions des sens et de l'esprit*, p. 198.
18 *Revue Philosophique* (1894), pp. 351–352.
19 *Journal de psychologie* (1904), pp. 17–27.
20 The idea of a resemblance of affective coloration would belong more particularly to M. Boirac, *Revue Philosophique* (1876), p. 431
21 Ribot and William James, who both thought out an explanation of this kind, were careful to add that they proposed it only as applicable to certain special cases: Ribot, *Les Maladies de la mémoire* (1881), p. 150; James, *Principles of Psychology* (1890), vol. I. p. 675.
22 *Archiv für Psychiatrie* (1887), pp. 409–436.
23 The hypothesis of M. Grasset, according to which the first experience had been registered by the unconscious, would, strictly speaking, avoid the last two objections, but not the first.
24 A.L. Wigan, *A New View of Insanity: The Duality of the Mind* (London, 1884), p. 85.
25 *Allgemeine Zeitschrift für Psychiatrie* (1868), pp. 48–63.
26 *Revue des Deux Mondes* (1885), p. 154.
27 *Archiv für Psychiatrie* (1878), pp 57–64.
28 *Revue Philosophique* (1902), pp. 160–163.
29 *Revue Philosophique* (1893), pp. 485–497.
30 *Annales medico-psychologiques* (1896), p. 455.
31 *Proceedings of the Society for Psychical Research* (1895), p. 343.
32 *Archives de psychologie* (1903), pp. 101–110.
33 *Revue Philosophique* (1894), pp. 34–35.
34 Ribot, *Les maladies de la mémoire*, p. 152.
35 *L'Illusion de fausse reconnaissance*, 1898. The reading of this book, which describes many new cases, is indispensable to the student of the subject. Mlle. J. Tobolowska, in her *Étude sur les illusions du temps des rêves* (1900), adopts M. Bernard-Leroy's conclusions.
36 Cf. *Matter and Memory*, pp 167–175 [*Œuvres*, pp. 306–314].
37 Janet, *Les Obsessions et la psychasthénie* (1903), vol. I. p. 287; also *Journal de psychologie* (1905), pp. 289–307.
38 *Revue de Psychiatrie* (1903), pp. 139–166.

Memory of the Present and False Recognition 63

39 *Zeitschrift für Psychologie* (1904), pp. 321–343.
40 *Journal de Psychologie* (1905), pp. 216–228.
41 In the same way, *depersonalisation* has been explained as a 'lowering of vital tone'. Cf. Dugas, '*Un cas de dépersonnalisation*', *Revue Philosophique* (1898), pp. 500–507.
42 See *Matter and Memory*, Chapter III, especially 172–175 [*Œuvres* pp. 311–313].
43 Bergson observes a more or less systematic distinction between *la mémoire* and *le souvenir*. Memory in the first sense denotes something like a faculty of the mind (strictly speaking, it would be more accurate to understand *la mémoire* ontologically, as Bergson does in *Matter and Memory*, namely as a 'virtual multiplicity'); in the second sense, it denotes either an event of recollection or that which is recollected by this faculty. We have chosen to translate both words as *memory* throughout, including the French in order to indicate whenever *la mémoire* is being used. [Editor's note]
44 See *Matter and Memory*, Chapter II, especially pp. 99ff; and pp. 148–9 [*Œuvres*, pp. 244ff, 289–291].
45 See *Matter and Memory*, pp. 127–131 [*Œuvres*, pp 270–275]; cf. the whole of the first chapter.
46 We translate *l'élan* throughout as 'impetus'. [Editor's note]
47 See, in particular, the cases collected by Bernard-Leroy, *op. cit.*, pp. 176, 182, 185, 232, etc.
48 Bernard-Leroy, *op. cit.*, p. 186.
49 Lalande, A., '*Des paramnésies*', *Revue Philosophique* XXXVI (1893), p. 487.
50 Jensen, *op. cit.*, p. 57.
51 F. Gregh, cited by Bernad-Leroy, *op. cit.*, p. 183.
52 Bergson is referring to the point he makes in *Matter and Memory*, pp. 163f. [*Œuvres*, pp. 302ff.]. The observation is situated within the context of a discussion of general ideas, in which he once again stresses the 'purely utilitarian origin of our perception of things.' [*Matter and Memory*, p. 158/*Œuvres*, p. 299] [Editor's note]
53 An observation recounted by M. Bernard-Leroy, *op. cit.*, p. 169.
54 *Zeitschrift für Psychologie*, vol. 36 (1904), pp. 321–343; and vol. 43 (1906), pp. 1–17.
55 See in particular the self-observations of Kräpelin and of Dromard and Albès, *art. cit.*

4
The Instant

Gaston Bachelard
Translated by Mary McAllester Jones

'Lovely, lively, virginal today'. (Mallarmé)
'We shall have lost even the memory of our encounter...
And yet we come together, only to separate and come together once more,
Where the dead come together: on the lips of those who live'. (Samuel Butler)[1]

I

Time has only one reality, the reality of the Instant: this is the metaphysical idea central to Gaston Roupnel's argument.[2] In other words, time's reality is compressed into the instant and suspended between nothingness and nothingness. And while time can doubtless be reborn, it must first die. Time cannot carry its being from one instant to another to make it a duration. The instant is already solitude... It is solitude at its metaphysically most stark. The instant's tragic isolation is indeed confirmed by solitude of a more emotional nature: through a kind of creative violence, time which is limited to the instant isolates us not just from others but from ourselves, breaking as it does with the past we hold most dear.

Thus, from the very outset of Roupnel's reflection – reflection on time being the task to be performed before embarking upon any metaphysics – he comes face to face with the idea that time presents itself to us as the solitary instant, as consciousness of solitude. We shall see in due course how the phantom of the past and the illusion of the future are reconstituted; however, if we are fully to understand Roupnel's *Siloë*, the complete equality of the present instant and reality must first of all be thoroughly absorbed. How can reality escape being marked by the present instant and vice versa, how can the present instant fail to set its stamp on reality? If my being is conscious of itself only in the present

The Instant

instant, how can we not see that the present instant is the sole domain in which reality is experienced? Were we to eliminate our being at some later point, we would indeed have to begin with ourselves in order to prove being. First then let us take our thought: we shall feel it ever fading away with each passing instant, leaving neither memory of what has just left us nor even hope – since there is no consciousness – for what the instant to come will bring. Gaston Roupnel writes that:

> It is of the present and only of the present that we are conscious. The instant that has just escaped us is the same vast death to which belong annihilated worlds and firmaments grown cold. And the same fearsome unknown holds within it, in the same darkness of the future, both the instant that is drawing closer to us and also Worlds and Heavens that have as yet no knowledge of themselves (*Siloë*, p. 108).

Roupnel adds a point that we shall contradict simply and solely in order to highlight his ideas once more: 'there are', he writes, 'no degrees in this death, which is as much the future as the past'. In order to reinforce the isolation of the instant, we would go as far as to say that there are indeed degrees in death and that what is more dead than death is what has just gone… Indeed, reflection on the instant persuades us that forgetting is most marked when the past it destroys is close, just as uncertainty is most moving when placed on the axis of thought that is to come, in the dream we seek yet that we feel already deceives us. A phantom with a little more coherence and solidity could perhaps come back from the remoter past and live again, through the effect of a wholly formal permanence which we shall be studying later on. However, the instant that has just passed cannot be retained as a complete being, in all its individuality. The recollection of many instants is required in order to make a complete memory. In the same way, the cruellest of bereavements is consciousness of a future that has been betrayed: when that heart-rending instant comes as the eyes of a beloved being closed, we feel at once the hostile newness with which the next instant 'assails' our heart.[3]

Its dramatic character may enable us to glimpse the reality of the instant. We would like to stress that in a rupture in being of this kind, the idea of discontinuity is clearly indisputable. It may be objected that these dramatic instants separate two more monotonous durations. However, we describe as monotonous and regular any development that we do not examine with eager attention. Were our hearts big enough to love life in all its detail, we would see that all instants both give and take away at one and the same time, and that newness – young or tragic but always coming suddenly – constantly shows us the essential discontinuity of Time.

II

Clearly though, this recognition that the instant is the primordial element of time cannot be conclusive until the ideas of the instant and of duration have first been compared. Inevitably then, readers will think back to Bergsonian theory here, even though Roupnel's *Siloë* bears no trace of polemics. Since we have set ourselves the task of recording in these pages all the thoughts of an attentive reader, we shall have to note down all the objections arising from our recollection of Bergson's discussions. Moreover, the intuition we are presenting here will perhaps best be understood if we set Roupnel's argument against Bergson's.

This then is the plan of our discussion. First, we shall remind readers of the essentials of the theory of duration and examine as clearly as we can both terms of the following antithesis: Bergson's philosophy is a philosophy of duration, Roupnel's is a philosophy of the instant. Second, we shall try to show our own attempts to reconcile these two. We shall not though espouse the theory intermediate between them that did attract us momentarily; if we go back over this again here it is simply because it strikes us as likely to occur to eclectic readers, and because it may also lead them to put off having to decide. Lastly, having described our own debates, we shall see that Roupnel's theory adopts what for us is the clearest and most prudent position, that corresponding to the most direct consciousness of time.

Let us therefore first examine Bergson's position.

According to Bergson, we experience duration directly and within ourselves. This duration is indeed an immediate datum of consciousness.[4] It may of course eventually be elaborated, objectified, and distorted. Physicists for example, devoted as they are to abstractions, manage to turn duration into uniform, lifeless time, time which has no end and no discontinuity. This completely dehumanised time they then put into mathematicians' hands. Once it has entered the realms of these prophets of abstraction, time is reduced to being simply an algebraic variable, the supreme variable in fact, which from then on is more suited to the analysis of possibility than to the study of reality. Indeed for mathematicians, continuity does not characterise a reality but is instead the schema of pure possibility.

What then is the instant for Bergson? He regards it as no more than an artificial break which helps geometricians to think schematically.

Because our intellect is unable to deal with living processes, it immobilises time in a present which is always artificial. This present is pure nothingness, unable even to establish any real separation between past and future. Indeed, it seems that the past carries into the future the forces within it; it seems too that the future is necessary in order to furnish an outlet for these forces within the past, and that one single and identical vital impetus gives duration its unity. Thought is but a fragment of life and it ought not to impose its rules on life. Intellect is wholly absorbed in its study of static being, of spatial being, and must therefore take care not to misunderstand the reality of becoming. Lastly, there is in Bergson's philosophy the indissoluble union of past and future. Time must therefore be grasped as a whole if it is to be grasped in its reality. Time lies at the very fount of the vital impetus. Life can be the recipient of instant illustrations, but it is in the end duration that explains life.

Having now reminded ourselves of Bergsonian intuition, let us consider where the problems lie.

First, it can be seen that Bergson's argument has repercussions for the reality of the instant.

Indeed if the instant is a false caesura, it will be very hard to distinguish past and future since their separation is always artificial. We must therefore take duration as an indestructible unity. From this stem all the consequences of Bergson's philosophy: in each act of ours, in our slightest movement, we should therefore be able to grasp in all its completeness what is then just taking shape, the end in the beginning, being and all its becoming in the burgeoning seed.

Let us suppose however that past and future can be merged together once and for all. This being so, it seems to us that there is a real problem here for anyone wishing to make full and complete use of Bergsonian intuition. Given such convincing proof that the instant is unreal, how are we to speak of the beginning of an act? What supernatural power, lying beyond duration, will enjoy the privilege of ascribing a decisive role to one particular fruitful hour which, if it is to have duration, must even so have a beginning? This theory of beginnings, the full significance of which we shall see in Roupnel's philosophy, cannot help but be obscure for a philosophy that is its contrary and that denies the value of the instantaneous. If we consider life in its middle course, growing and developing, then we shall doubtless be able to show, along with Bergson, that the words *before* and *after* are to be understood as reference points and nothing more, because between past and future we pursue a process of development which by virtue of its general success appears continuous. If however we venture into the realm of sudden, swift mutations, where the creative act takes place suddenly, swiftly, how can we not see that a new era is always inaugurated by an absolute? But all evolution, to the extent that it is decisive, is punctuated by creative instants.

Where can we be more sure of finding this knowledge of the creative instant than in the springing-up of our own consciousness? Surely it is here that the vital impetus is most active. What is the point of trying to go back to some mysterious hidden force which has more or less failed to gather its own momentum, unable either to realise it fully or even to *continue* it, when we can see before us in the active present the thousand accidents of our own intellectual development, the thousand attempts to renew and to create ourselves? Let us go back then to the starting-point of idealist philosophy and agree to take as our experimental field our own mind as it strives towards knowledge. Knowledge is above all else the work of time. Let us try then to free this mind of ours from the bonds of the flesh and from the prison-house of matter. Once it is set free we realise that, depending on the degree of freedom given, the mind receives a thousand incident events and that the line of the mind's dream will be refracted, broken into a thousand segments that hang suspended from a thousand vertices.[5] As the mind pursues and creates knowledge, we see that it is a sequence of clearly separate instants. It is in writing the history of the mind that psychologists, like all other historians, introduce quite artificially the bond of duration. Deep within ourselves, where gratuitousness has so clear a meaning, we cannot grasp that causality which would give duration its force: the search for causes in a mind in which only ideas are born is, in our view, a purely academic and peripheral problem.

Let us sum up: whatever we may think of duration in itself, as it is grasped in Bergson's intuition of it which we do not claim to have invalidated in these few pages, we must at the very least concede that, along with duration, the *instant* is both real and decisive.

We shall moreover have the opportunity to return later on to the argument against the theory of duration as an *immediate datum* of consciousness. With the help of Roupnel's intuitions, we shall then show how duration can be constructed with instants that have no duration. This will, in our view, be proof positive that the instant is metaphysically primordial in character and that, as a result, duration is indirect and mediate.

We are keen though to return to positive discussion. In any case, Bergson's method permits us from now on to make use of psychological observation. We must therefore conclude as Roupnel does that: 'the Idea

The Instant

we have of the present is singularly complete, evident, and positive. Our entire personality is grounded here, and here alone – both through and in the present – we have the sensation of existence. And between the feeling of the present and that of life, there is absolute identity' (*Siloë*, p.108). From the standpoint of life itself, we must therefore try to understand the past by means of the present instead of endlessly striving to explain the present in terms of the past. Doubtless the sensation of duration will eventually have to be clarified. Let us for the time being take it to be a fact: duration is a sensation like any other and as complex as all sensations are. And let us not shrink from drawing attention to its apparently contradictory nature: duration is made up of instants without duration just as a straight line is made up of points without dimension. Basically, if entities are to contradict each other, they have to be active in the same zone of being. If we establish that duration is a relative and secondary datum which is always more or less artificial, then how could the illusion we have of it contradict our immediate experience of the *instant*? All these reservations are expressed here so as to prevent us being accused of an outright vicious circle when in fact we are taking these words in their general sense, without any reference to their technical meaning. Having taken these precautions, we can state as Roupnel does that:

> Our acts of attention are episodes of sensation extracted from the continuity we call duration. Yet the continuous fabric on which our minds embroider discontinuous patterns of acts is but the mind's laborious and artificial construct. Nothing permits us to maintain that there is duration. Everything in us contradicts its meaning and destroys its logic. Moreover, our instinct is better informed here than is our reason. The feeling we have of the past is one of negation and destruction. Credit granted by the mind to so-called duration, duration that no longer is and in which the mind no longer is, cannot be repaid since duration's account is empty (*Siloë*, p. 109).

We must in passing emphasise the place the act of attention has in our experience of the *instant*. This is because the only clear and indisputable fact is indeed the will, the consciousness that tenses itself until it decides to act.

The action taking place behind the act is already ranked among consequences that are logically or physically passive. Here then we have an important nuance which distinguishes Roupnel's philosophy from Bergson's: *Bergson's philosophy is a philosophy of action; Roupnel's is a philosophy of the act*. For Bergson, an action is always a continuous sequence which places between the decision and the goal – each of them more or less schematic – a duration that is always original and real. For those who favour Roupnel, an act is first and foremost an instantaneous decision and it is this decision that is responsible for originality. Speaking now in terms of physics, the fact that in mechanics an impulse is always the compound of two different infinitesimal orders leads us to compress the deciding, initiating instant down to its very limits as a point. An impact, for instance, is understood as an infinitely large force acting in an infinitely short time. It would be possible moreover to analyse the sequence of events following on from a decision in terms of what are in fact subordinate decisions. It would be seen that a varying movement – the only movement that Bergson quite rightly considers to be real – continues by following the same principles that start it in motion. It becomes increasingly difficult to observe discontinuities in the sequence as progressively the action following on from the act is consigned to less conscious organic reflexes. It is for this reason that we have to go back to clear acts of consciousness if we are to feel the instant.

When we come to the last pages of this essay we shall, if we are to understand the relationship between time and progress, need to return to this actual and active conception of the experience of the instant. We shall then see that life cannot be understood in passive contemplation; understanding life involves not just living it but in fact propelling it forward. Life does not flow down a slope, along the axis of an objective time that, as it were, serves as a channel for it. And while it is a form imposed on time's line of instants, life's prime reality lies always in the instant. Consequently, if we go to the root of psychological certainty, to the point at which sensation is no more than an ever complex reflection of or response to an ever simple voluntary act, when attention is condensed, compressing life into a single isolated element, then we realise that the instant is what is truly specific to time. The deeper we take our meditation on time, the more this meditation shrinks. Inactivity is the only thing that lasts: acts are instantaneous. Can we not say then that conversely the instantaneous is an act? Take an idea with very little to it and then compress it into an instant: see, it enlightens our mind. The repose of being is, on the contrary, already nothingness.[6]

How therefore can we fail to see that, in a striking verbal concurrence, the nature of the act is to be actual? And how can we not then see that life is the discontinuity of acts? It is this intuition that Roupnel presents to us particularly clearly in the following passage:

The Instant

It has been said that duration is life. This may be so, but life must at the very least be set within the framework of the *discontinuity* containing it and also within the form revealing it, a form which assails. Life is no longer that fluid continuity of organic phenomena which all flowed into one another, merging together in functional unity. Being – that strange site where material memories are commemorated – is for itself simply a habit. What can be permanent in being is the expression not of a constant and immobile cause but rather of a juxtaposition of results that are both fleeting and incessant, each with its own separate basis, results which, when bound together by what is simply habit, make up an individual (*Siloë*, p. 109).

Bergson no doubt had to ignore accidents when writing his epic account of evolution. Roupnel, working with a historian's attention to detail, could not fail to see that every action, however simple, must of necessity break up the continuity of life's becoming. If we look at the history of life in its detail, we realise that it is like any other history, full of repetition and anachronism, of things attempted, of failure and fresh starts. The only accidents to which Bergson paid attention were those revolutionary acts in which the vital impetus split itself up and the genealogical tree divided, branching in different directions. He did not need to draw in detail when composing a vast fresco of this kind. In other words, he did not need to draw objects. He therefore came in the end to produce that impressionist painting which we see in *Creative Evolution*. Illustrating as it does an intuition, this is not so much a portrait of things but rather the image of a soul.

However, philosophers who wish to go into every atom, every cell, and every thought in their description of the history of things, of living beings, and of the mind must reach the point when they have to separate facts from each other, facts being acts and acts, even when incomplete or botched, have at least necessarily to *begin* in the absolute of birth. Effective history must therefore be described with beginnings; we must follow Roupnel and produce a theory of *accidents as principle*.

There is only one general law in truly creative evolution, the law that an accident is at the root of any attempt at evolving.

We see then that Roupnel's intuition of time is the exact opposite of Bergson's in its consequences with regard to the evolution of life as well as in its first intuitive form. Before we go any further, let us give the following schematic summary of the difference between these two theories.

For Bergson, time's true reality is duration; the instant is only an abstraction and has no reality. The instant is imposed from the outside by intellect, which can only understand becoming by locating motionless states. We can therefore represent Bergsonian time fairly accurately by a straight black line on which we place a white dot, symbolising the instant as nothingness, as a fictitious void.

For Roupnel, time's true reality is the instant; duration is only a construction and has no absolute reality. It is made from the outside by memory, that supreme example of our powers of imagination, which wants to dream and relive but not to understand. We can thus represent Roupnelian time fairly accurately by a straight white line, entirely potential and entirely possible, on which like some unforeseeable accident a black dot suddenly sets itself, symbolising an opaque reality.

It must be noted moreover that this linear disposition of instants remains for Roupnel as for Bergson an artifice of imagination. Bergson sees this duration that is unfolded in space as an indirect means of measuring time. Yet length of time does not represent the value of duration, and we would have to come back up from extensible time to intensive duration. Here again, the argument for discontinuity can be adapted without difficulty: intensity can be analysed by the number of instants in which the will is clear and tensed in readiness, every bit as easily as by the gradual, fluid enrichment of the self.[7]

Let us now open a parenthesis before we go more specifically into the argument presented in *Siloë*.

III

We have already said earlier on here that we ourselves long hesitated between these two intuitions, seeking in fact to follow the path of reconciliation and bring together the advantages of both theories in a single schema. We did not in the end find satisfaction in this eclectic ideal. However, since we have set ourselves the task of studying the effect on us of intuitive reactions derived from the intuitions outlined above, we owe our readers a detailed account of our failure.

We would have liked first of all to give the instant a dimension and make of it a kind of atom of time which would retain within itself a certain duration. We told ourselves that an isolated event must have a short, logical history regarding itself, in the absolute of its own internal development. We well understood that its beginning could be relative to an accident that was external in origin; but in order for it to shine and then decline and die, we asked that being, however isolated, should be given its share of time. We accepted that the ideal of life is the life ardently lived by a mayfly, but we required that between dawn and its nuptial flight the mayfly should be able to enjoy a rich inner life. We therefore still wanted duration to be a deep and immediate richness of being. This then was our first position regarding the *instant*, which would then have been a small fragment of Bergsonian continuity.

Let us now turn to what we then took from Roupnelian time. We imagined that the atoms of time could not touch each other or rather, that they could not merge into one another. What would always stop this merging is the inalienable newness of instants, of which the theory of accidents found in *Siloë* had convinced us. In a theory of substance, which is in fact not far from being tautological, qualities and memories will be carried from one instant to another without difficulty; we shall never manage to make the permanent explain becoming. If therefore newness is essential to becoming, we have everything to gain by attributing this newness to Time itself: it is not being which is new in time that is uniform but rather it is the instant which, as it renews itself, carries being forward to freedom, to the very first step on the path of becoming. Moreover by virtue of its attack, the instant makes its presence felt suddenly and in all its entirety; it is the agent of the synthesis of being. According to this theory, the instant therefore necessarily reserves its individuality. As for the problem of knowing whether atoms of time touched each other or were separated from each other by nothingness, this seemed to us a secondary issue. Or rather, once we accepted the constitution of atoms of time, we came to think each of them in isolation and for the sake of this intuition's metaphysical clarity, we realised that a void was necessary – whether it in fact existed or not – in order for us to imagine an atom of time correctly. It therefore struck us as useful to condense time round *nuclei of action* in which being found itself again in part, while at the same time drawing from the mystery of Siloam the inventiveness and energy needed in order to become and to progress.

Lastly, by bringing these two theories together we therefore reached a fragmented Bergsonism, a vital impetus broken up into impulses, and temporal pluralism which, accepting as it does diverse durations and individual times, seemed to us to offer means of analysis as flexible as they are rich.

Yet metaphysical intuitions constructed on an eclectic ideal very rarely endure. A fruitful intuition must first demonstrate its unity. It was not long before we realised that in reconciling them, we had combined the difficulties of both theories. A choice had to be made, not when we reached the end of our deliberations but at the very foundation of these intuitions.

We shall now explain then how we went from the *atomisation of time*, which had been our stopping-point, to the absolute *temporal arithmetisation* for which Roupnel consistently argues.

Firstly, it was a false conception of the order of metaphysical entities that had beguiled us and led us into the blind alley in which we have just been getting nowhere very much: by keeping in contact with Bergsonian theory, we wanted to put duration into the very space of time. Without any argument, we considered this duration as time's only quality and as a synonym of time. Let us now admit that this duration is simply a postulate. We can only judge the value of this postulate according to the clarity and import of the construction it favours. However, we always have the *a priori* right to start from a different postulate and to try out a new construction in which duration is deduced rather than postulated.

Yet this *a priori* consideration would not of course have been sufficient to take us back to Roupnel's intuition. Indeed, supporting Bergson's conception of duration there remained all the many proofs he has marshalled of the objectivity of duration. Bergson undoubtedly asked us to feel duration within us, in our personal experience of our own innermost self. He went further than this, however. Bergson showed us objectively that we are all part of one single surging force, all swept along by the same tide. Should boredom or impatience make an hour last longer or happiness shorten our day, then impersonal life, the life of other people, will remind us of the real nature of Duration. We have only to watch a simple experiment, a lump of sugar dissolving in a glass of water, and we shall realise that there is indeed an objective, absolute duration that corresponds to the duration we ourselves feel.[8] This being so, Bergsonian theory claimed that here it re-entered the realm of measurement while still preserving all that we know through our innermost intuition. It argued that while our soul is in immediate contact with the temporal quality of being and with the essence of its becoming, the objectivity of becoming belongs to the realm of time as a quantity, however indirect our means of examining this might be. It seemed then that both discursive proof and the evidence of intuition safeguarded the primary character of Duration.

Let us now explain how our own confidence in Bergson's argument was shaken.

We were aroused from our dogmatic slumbers by Einstein's criticism of objective duration.

It was very soon clear to us that Einstein destroyed the absoluteness of that which has duration, while maintaining, as we shall see, the absolute character of what is, that is to say the absolute character of the instant.

It is the *lapse* of time, the 'length' of time that becomes relative in Einstein's theory. Length of time is shown to be relative to the method of measuring it. It is said that, were we to take a trip into space at a fairly high velocity, we should find on our return that the earth was a few centuries older while according to the clock we took with us, only a few hours would have passed. A much shorter journey would be required to adjust the time it takes for a sugar lump to dissolve in a glass of water, a time which Bergson postulates as fixed and necessary, so that it becomes synchronous with our impatience as we watch and wait.

We are not just playing pointless mathematical games here, and we would emphasise this straightaway. The relativity of the lapse of time in systems that are in motion must now be regarded as a scientific fact. Were it considered legitimate to reject the lessons of science in this respect, we should have to be given the right to entertain doubts concerning the intervention of physical circumstances in the sugar-dissolving experiment, and doubts too as to whether time really does interfere with experimental variables. For instance, do we not all accept that temperature is involved in this experiment with the sugar lump? Very well then, let us say that for modern science this experiment also involves the relativity of time. We have to accept science as a whole and not try to limit it to just some of its aspects, beyond which we refuse to go.

In this way then, Relativity brought the sudden ruin of everything to do with external proofs of one unique Duration, that fundamental principle ordering all events. Metaphysicians had to retreat into their own local time, shutting themselves into their own innermost duration. Immediately at least, the world offered no guarantee that our own individual duration, the duration that each of us lives within our own innermost consciousness, would ever converge with the duration lived by another person.

We now come to something to which it is well worth paying close attention: *the instant, clearly and accurately defined, remains in Einstein's theory an absolute*. We can give this absolute value to the instant simply by considering it in its synthetic state, as a point in space-time. In other words, we must take being as a synthesis comprising both space and time. It lies at the point of concurrence of place and present time: *hic et nunc*, not here and tomorrow, not there and today. In these last two expressions, the point would expand along either the axis of durations or an axis of space; there is in each of these expressions something that makes a precise synthesis impossible, both leading therefore to an entirely relative study of duration and space. Yet once we agree to weld and fuse together the two adverbs here and now, *hic et nunc*, the verb 'to be' will at last receive its full force as an absolute.

In this very place and at this very moment: here, simultaneity is unmistakable, it is evident and precise; here, succession is ordered without any hesitation or any obscurity. One consequence of Einstein's theory is that we can no longer claim that the simultaneity of two events localised at different points in space is self-evident. For this simultaneity to be established, we should need an experiment in which we could be sure, beyond a shadow of doubt, that the stationary ether exists. Michelson's failure put an end to all hopes of ever carrying out such an experiment. We must therefore arrive at an *indirect* way of defining simultaneity in different places and, as a result of this, we shall have to adjust our measurement of the duration separating different instants in conformity with this new and always relative definition of simultaneity. There can be no true concomitance without coincidence.

We return then from our foray into the realm of the phenomenon, convinced that duration only accumulates in an artificial way, in a climate of pre-existing conventions and preliminary definitions, convinced also that its unity derives from nothing other than the generality and sluggishness of our own investigation. Conversely, the instant is shown to be capable of precision and objectivity, and we sense here in the instant the mark of something fixed and absolute.

Are we now going to make the *instant* into the centre of condensation around which would be placed a duration that fades and disappears? In other words, are we now going to give the instant just sufficient continuity to make an isolated atom of time, standing out in relief against a background of nothingness? Will it be given sufficient continuity too for Nothingness to be engraved, like a cameo, with its two deceiving faces, seen when we either look back to the past or turn towards the future?

This was the last solution we attempted before we came in the end, without compromising on anything at all, to accept Roupnel's very distinct standpoint.

Let us explain what brought about our final conversion.

When we still accepted Bergson's idea of duration, we set out to study it by trying very hard to purify and consequently impoverish duration as it is given to us. Yet our efforts would always encounter the same obstacle since we never managed to overcome the lavish heterogeneity of duration. We laid the blame of course on our own inept meditation and on our failure to detach ourselves from the continual onslaught of all that is accidental and new. Never at any time did we manage to lose ourselves so completely that we then truly found ourselves, nor did we ever meet with success in our efforts to reach and then follow something in us that is uniform and flowing, where duration slowly sets before us a life story in which nothing is lived, a happening in which nothing happens. We would have preferred it if becoming were like a flight in a cloudless sky, disturbing nothing and hindered by nothing, soaring higher and ever higher into emptiness; in short, we would have preferred to find becoming in all its solitude, in all its purity and simplicity. Many a time have we sought to discover elements in becoming that would be as coherent and as clear as those Spinoza found in his meditation on being.

We remained, even so, quite incapable of finding these endless, unbroken lines within us, the simple sweeping lines traced by the vital impetus in the picture it draws of becoming. In due course, as one might expect, we tried to find this homogeneous character of duration by confining our study to smaller and smaller fragments. Yet we were still dogged by failure. There was far more to duration than just having duration, for duration was truly alive! However small the fragment under consideration, we had only to examine it microscopically to see in it a multiplicity of events; always it was the embroidery that we saw, never the fabric, always the shadows and reflections mirrored on a restless river, never its deep, pellucid waters. Duration, like substance, can offer us only phantoms, only shadowy things. Indeed, duration and substance endlessly enact the fable of the deceiver deceived, the one deceiving the other only to be in its turn inevitably deceived, for becoming is the phenomenon of substance and substance is the phenomenon of becoming.

Should we not agree then that it is metaphysically more prudent to equate time with what is accidental, which would mean in fact that time is to be equated with its phenomenon? Time can be observed solely in instants; duration, as we shall see, can be experienced solely in instants. Duration is scattered in a cloud of instants or better, it is a group of points which are drawn more or less closely together by a *phenomenon* of perspective.[9]

We have indeed become aware of the need to descend now to where we find temporal points with no individual dimensions. The line joining the points together and schematising duration has only a panoramic and retrospective function, the indirect and secondary subjective character of which we shall subsequently show.

Although we do not wish to go into psychological proofs at any length, let us just point out here the psychological nature of the problem. It needs to be recognised that the *immediate* experience of time is not the very fleeting, very difficult and academic experience of duration but rather the nonchalant experience of the instant that is always grasped as immobile. Everything in us that is simple and strong, and even everything in us that endures, is a gift an instant brings.

We shall start by grappling with the most difficult aspects of this problem, with for example the important point that memory of duration is one of the least enduring memories. We remember that we have been, we do not though remember that we have had duration. Distance in time distorts the perspective of length since duration is always dependent on a point of view. Besides this, pure memory in Bergson's philosophy is nothing other than an image taken in isolation. Had we the opportunity in a longer work to examine the problem of how memories are localised in time, we would have no difficulty showing how ill situated they are and how artificial is the order they find in our own inward history. The whole of Halbwachs's splendid book on the social frameworks of memory[10] would prove to us that our meditation does not in any way have at its disposal a solid psychological structure, the skeleton of dead duration in fact, in which naturally, psychologically, and in the solitude of our own consciousness we could firmly place the memory we have recalled. When it comes down to it, we need to learn and relearn our own chronology and we do this with the help of synoptic tables, summing up as they do the most accidental of coincidences. Thus it is that the history of kings is recorded in the humblest of hearts. We would have a poor idea of our own history or at least it would be full of anachronisms if we paid less attention to contemporary History. An election with as little importance as that of a President of France[11] allows us to place a personal, private memory of ours both rapidly and accurately, proving surely that we have not preserved the least little trace of durations that are now dead and gone. Memory, that guardian of time, keeps only the instant; it preserves nothing, absolutely nothing at all, of that complicated, artificial sensation of ours that is duration.

In the same way, the psychology of will and of attention – that will of our intellect – leads us to take as our working hypothesis Roupnel's conception of instants without duration. It is by now a matter of certainty that in this psychology, duration can only intervene indirectly; we can see fairly easily that it is not a primordial condition. With duration, we can perhaps measure the period in which we wait for something to happen but we cannot measure our attention to that thing, because the intensity which characterises attention is given in a single instant.

This problem of attention arose of course in our own meditations on duration. Unable ourselves to keep our attention fixed for any length of time on that ideal of nothing, the self stripped of everything and laid bare, we were in fact tempted to break up duration according to the rhythm of our acts of attention. Here again as we confronted the very minimum of unexpectedness in our attempt to rediscover the realms of pure inwardness, of inwardness stripped of all else, we suddenly realised that this attention to ourselves brought with it, by virtue of the way it worked, all the fragile and delightful newness of thought that has no history, of thought without thoughts. This kind of thought, compressed as it is in its entirety into the Cartesian cogito, has no duration. It is evident simply because it is instantaneous; it becomes fully conscious of itself simply because it is empty and solitary. It then awaits the world's attack, in a duration that is but a nothingness of thought and real nothingness consequently. The world brings it knowledge, and it is once again in a fruitful instant that our attentive consciousness will be enriched with objective knowledge.[12]

Moreover, since attention has both the need and the ability to go and start tackling things all over again, it is in essence entirely constituted by these fresh starts. Attention too is a series of beginnings; it is made up of the renewals of a mind coming back to consciousness as time marks its instants. In addition, if we extended our study to that very small area in which attention becomes decision, we would see the flashes and fulgurations of a will in which clarity of motive and the joy of acting converge. We could then speak of truly instantaneous conditions. Such conditions are strictly preliminary or better, they are pre-initial since they are antecedent to what geometricians call the initial conditions of movement. This is why they are metaphysically and not abstractly instantaneous. If you watch a cat stalking its prey, you will see the *instant of evil* making its mark in reality; Bergsonians though will always end up considering the trajectory of evil, however compressed their study of duration may be. The pouncing cat as it begins its leap no doubt initiates the sequence of a duration in accordance with physical and physiological laws, laws that govern complex wholes. Yet before the complicated process of the springing forward, there was the simple, criminal instant of decision.

Besides this, if we direct this attention to what is going on around us, making it attention to life instead of attention to our personal and private thought, we realise at once that its starting-point is always a coincidence. Coincidence is the very minimum of newness necessary in order to fix our minds. We cannot pay attention to a process of development in which duration is the only principle ordering events and differentiating between them. Newness is needed if thought is to intervene; newness is needed too if consciousness is to assert itself and if life is to progress. And newness is of course, in its very principle, plainly always instantaneous.

Finally, the psychology of the will, of certainty, and of attention could best be analysed by the point of space-time. Unfortunately though, such an analysis can only become clear and convincing once philosophical language and even ordinary language have assimilated the theories of Relativity. We are aware in fact that this process of assimilation has begun, and yet it is far from complete. We believe however that this is the way to bring about the fusion of both spatial and temporal atomism. The deeper we take this fusion, the more we shall understand the importance of Roupnel's argument. It is in this way that its concrete character will best be grasped. The space-time-consciousness complex is atomism with a triple essence; it is the monad in its triple solitude, communicating with neither things, nor with the past, nor with other souls.

Yet all these presumptions will seem all the more flimsy since many habits of thought and language oppose them. We are moreover well aware that our argument will not carry conviction straightaway and that psychology may strike many readers as an inappropriate field in which to pursue metaphysical research.

What did we hope to achieve by bringing all these ideas together? Our aim was simply to show that we would, if necessary, agree to fight on the most unfavourable ground. In the end though the most important thing is to pose the problem in metaphysical terms and it is to this that we shall now direct our energies. Let us therefore take the argument in all its clarity. Roupnel's intuition of time affirms two things, firstly that time is absolutely discontinuous, and secondly that the instant is absolutely punctiform.

Roupnel's argument brings about then the most complete and unequivocal arithmetisation of time. Duration is simply a number whose unit is the instant.

For the sake of greater clarity, let us as a corollary negate the truly temporal and immediate nature of duration. According to Roupnel, 'Space and Time only appear infinite to us when they do not exist' (*Siloë*, p. 126). Bacon had already observed that 'nothing is more vast than empty things'. With these phrases as our inspiration, we can say without, in our view, distorting Roupnel's thinking, that in fact there is only *nothingness that is continuous*.

IV

We know only too well the kind of response we shall get to this last phrase. We shall be told that nothingness of time is in actual fact the interval separating instants which events mark out. And in order to give us a more thorough drubbing our opponents will, if they have to, accept that events are brought to birth instantaneously and that they are even, if need be, instantaneous. However, in order to distinguish between instants, they will require there to be an interval with real existence. They would like us to say that this interval is time, in fact, time that is empty and without events, time that has duration, duration that goes on and that can be measured. Yet we shall stubbornly insist that time is nothing if nothing happens in it, and that to speak of Eternity before creation does not make sense; we shall insist too that nothingness cannot be measured, that it cannot have magnitude.

While our intuition of wholly arithmetised time doubtless runs counter to what is commonly argued and may therefore clash with common ideas, this intuition ought to be judged for itself. It may seem to have very little to it and yet it should be recognised that up until now this intuition has, in all its developments, been coherent with itself.

If moreover we can bring in a principle that provides the basis for a substitute for the measurement of time, we shall in our view have finally overcome our opponents.

Let us sum up their criticism as bluntly as possible.

In your argument, they will say, you can no more accept the measurement of time than you can the division of time into aliquot parts,[13] and yet like everyone else you say that an hour lasts for sixty minutes and a minute lasts for sixty seconds. You believe therefore in duration. You cannot speak without using all those adverbs and all those words that suggest something that has duration, that passes, and that we expect. Even in this discussion, you are obliged to say things like 'for a long time', 'for such-and-such a period', 'during'. Duration is in grammar and morphology as well as being in syntax.

Yes of course, words are there before thought, before our attempts to renew thought. We have to use them as they are. Is it not however the philosopher's function to distort the meaning of words just sufficiently to draw the abstract from the concrete, allowing thought to escape from things? Ought not the philosopher's task be the same as that ascribed by Mallarmé to the poet, namely 'to give a purer sense to the language of the tribe'?[14] Just pause and reflect on the fact that because they are in part rooted in spatial aspects, all words expressing temporal characteristics involve metaphors and it will be seen that we have not been left crushed and defenceless on the battlefield of argument: those criticising us for a verbal vicious circle will no doubt withdraw their accusation.

The problem of measurement remains, however, and it is obviously here that criticism does seem decisive: since we *measure* duration it must have magnitude. It therefore bears the clear sign of its reality.

Let us therefore see whether this sign really is immediate. Let us try to show how, in our view, Roupnel's intuition understands duration.

What is it then that gives time its apparent continuity? It is the fact that by imposing a break *where we want it*, we seem to be able to point to a phenomenon which illustrates the instant that has been arbitrarily pointed out. We would thus be sure that our act of knowledge is open to full and free scrutiny. In other words, we maintain that our acts of freedom are placed on a continuous line *because we can at any moment always experience the efficacy of our acts*. We are sure of all this, but we are sure of this alone.

We shall now express the same idea using slightly different language which indeed must, at first sight, seem synonymous with that used above. We shall say then that *we can at all the times we wish experience the efficacy of our acts*.

This is where an objection arises. Does not the first way of expressing ourselves tacitly assume the continuity of our being? Is it not this continuity, assumed as being self-evident, that we then ascribe to duration? Yet what guarantee do we therefore have of the continuity that is attributed to ourselves in this way? The rhythm of our disconnected being need only correspond to *one* rhythm of the Cosmos for our examination to succeed every time. To put it more simply, all that is needed in order to prove the arbitrariness of the break we have made is for the *opportunity* for our own innermost act to correspond to an opportunity in the universe, in short for a coincidence to occur at a point in space-time-consciousness. Because of this, and this brings us to our main point, we consider *at all times* to be exactly synonymous in the argument in favour of the discontinuity of time with the word *always* in that in favour of temporal continuity. If this translation is allowed, then the whole language of continuity is restored to us through the use of this key.

Besides this, life puts at our disposal such a prodigious wealth of instants that, in terms of the account we take of them, this wealth indeed seems endless. We realise we could spend a great deal more, hence the belief that we could spend without counting. This is the source of our impression of inward continuity.

The Instant

Once we have understood the importance of a concomitance that is expressed by an agreement of instants, synchronism then becomes an obvious way of interpreting things in the hypothesis of Roupnelian discontinuity. Here again, a comparison should be made between Bergson's intuitions and those of Roupnel. For Bergsonian philosophers, two phenomena are synchronous if they are always in agreement. It is a matter of bringing together becoming and action. For Roupnelians, two phenomena are synchronous if all the times that one is present, the second is also there. It is a matter of bringing together fresh starts and acts.

Which expression is more prudent? Saying with Bergson that synchronism corresponds to two parallel sequences of events means we are going a bit beyond objective proofs and enlarging the field of our verification. We do not accept the metaphysical extrapolation that argues for a *continuum in itself* when we are always faced with only the discontinuity of our experience. Synchronism always appears therefore in a concordant numeration of effective instants; it never appears as what is in some ways a geometric measurement of continuous duration.

At this point, we shall doubtless be stopped in our tracks by a further objection. We shall be told that, even accepting that the phenomenon as a whole can be examined following the precise temporal schema of a cinematic shot, we cannot deny that dividing up time is always in fact possible and indeed always desirable if one wishes to follow the phenomenon's development in its every twist and turn; some ultra-refined film-making apparatus will then be cited that can describe becoming in ten thousandths of a second. Why then should we have come to a halt in our division of time?

The reason why our opponents postulate the endless division of time is because they always make their study at the level of a life as a whole, life summed up in the curve of the vital impetus. Since we live a duration which seems continuous in a macroscopic examination, we are led in an examination of detail to grasp duration in ever smaller fractions of our chosen units.

However, the problem would radically change if we considered how time is really constructed starting from instants, instead of how it is always artificially divided up starting from duration. We would then see that time is multiplied according to the schema of numerical correspondences, and that in no way is it divided according to the schema of the splitting up of a continuum.

The word 'fraction' is moreover already ambiguous. In our view, the theory of fractions as summed up by Couturat[15] should be mentioned here. A fraction is the grouping of two whole numbers in which the denominator does not really divide the numerator. Those in favour of temporal continuity differ from us in the following way with regard to this arithmetical aspect of the problem: our opponents start from the numerator which, for the needs of analysis, they consider to be a homogeneous, continuous quantity, and above all as a quantity that is immediately given; they divide this 'given' by the denominator, which is thus consigned to the arbitrariness of the examination, arbitrariness that increases as the examination becomes more detailed; our opponents might even be afraid of 'dissolving' duration were they to take infinitesimal analysis too far.

We on the other hand start from the denominator which is the mark of the phenomenon's richness in instants, the phenomenon being the basis of the comparison and known naturally with the greatest finesse: we maintain, in fact, that it would be absurd for there to be less finesse in the apparatus of measurement than in the phenomenon to be measured. With this basis as our reference point, we then ask ourselves how often there corresponds to this finely scanned phenomenon an actualisation of one that is more sluggish and inactive; the successes of synchronism give us at last the fraction's numerator.

Although the two fractions which are constituted like this may well have the same value, they are not in fact constructed in the same way.

We are of course aware of the following tacit objection: in order to determine the number of successes, do we not need some mysterious conductor of an orchestra who beats time both beyond and above the two rhythms being compared?[16] In other words, is it not to be feared, critics will say, that this analysis does tacitly use the word we have not uttered, the word 'during'? The real difficulty with Roupnel's argument, in fact, is avoiding the use of words taken from the ordinary, everyday psychology of duration. Yet once again, if we do practise reflection by going from a phenomenon rich in instants to one that is poor in them – from the denominator to the numerator – and not the reverse, we realise that we can manage not just without words evoking the idea of duration, which would be a simply verbal success, but also without the idea of duration itself: we would have demonstrated that here, where duration used to rule, its only use now is as a servant.

For greater clarity though let us offer a schema for this correspondence; let us then read this schema in two ways, one reading being in the language of duration and one in the language of instants, *both readings being made moreover from the perspective of Roupnel's argument.*

The Instant

Let us suppose that the macroscopic phenomenon be represented by this first line of points:

1. .   .   .   .

We set down these points without regard for the interval between them because, in our view, this is not where duration takes its meaning or its schema, and also because we consider the continuous interval to be nothingness, and nothingness has of course no more 'length' than it has duration.

Let us suppose that the more finely scanned phenomenon be represented, always with the same reservations being made, by this second line of points:

2. ...   .

Let us compare these two schemas.

If we now read like those in favour of continuity, from top to bottom – which is however a Roupnelian reading – we shall say that *while* phenomenon 1 occurs once, phenomenon 2 occurs three times. We shall bring in a duration dominating these series, a duration in which our word 'while' acquires its meaning. This duration will become clearer in ever more crude domains such as those of the minute, the hour, and the day...

If on the other hand we read synchronism like those unconditionally in favour of discontinuity, from bottom to top, we shall say that *every one in three times* there corresponds to phenomena having many apparitions (the phenomena which are the closest to real time), a phenomenon that has macroscopic time.

These two readings are in the end equivalent, but the first is a little too figurative; the second reading is closer to the original text.

Let us use a metaphor in order to explain what we mean. While the orchestra of the World has instruments which often fall silent, it is incorrect to say that there is always an instrument being played. The World turns in conformity with a musical beat imposed by the cadence of instants. Were we able to hear all the instants of reality, we would understand that it is not a case of the quaver being made up of bits of a minim but rather of the minim *repeating* the quaver. It is this repetition that gives rise to the impression of continuity.

Because of this, we understand that a relative wealth of instants will give us a kind of relative measurement of time. Should we wish to count up our temporal fortune very accurately, in short to be able to measure everything in us that is repeated, we would really have to live all Time's instants. It is in this totality that we would see the real way in which discontinuous time unfolds; it is in the monotony of repetition that we would rediscover the impression of empty and therefore pure duration. Taking a numerical comparison with the totality of instants as its basis, the concept of the temporal wealth of a particular life or a particular phenomenon would then acquire an absolute meaning according to how this wealth is used, or rather according to how it fails to be realised. This absolute basis is refused us, however, and we have to be content with just relative assessments.

We are now arriving, therefore, at a conception of duration as wealth which will serve the same ends as that of duration as extension. It can be seen to account not just for the facts but especially for the illusions, which is extremely important from a psychological point of view since the life of our mind is illusion before it is thought. We also understand that our constant illusions, illusions we come back to again and again, are no longer pure illusion, and that it is by reflecting on our errors that we come closer to truth. La Fontaine was right when he wrote of illusions 'which, always lying to us, never indeed deceive'.

The severity and rigour of rather academic kinds of metaphysics can be relaxed then, and we can return to the banks of Siloam where mind and heart are reconciled, each completing the other as they come together there. Duration's affective character, the joy or the pain of being, comes from the proportion or disproportion of the hours of life used as hours of thought or friendship. Matter forgets to be, life forgets to live, the heart forgets to love. It is when we sleep that Paradise is lost. Let us look too at the effects of our inactivity: an atom radiates and often exists; it uses a great number of instants but not however all instants. A living cell is already more sparing of its efforts, using only a fraction of the temporal possibilities offered by all the atoms that constitute it. And as for thought, it is in irregular bursts and flashes that thought uses life. Here we have three kinds of filtering which allow all too few instants to come to consciousness! We thus feel a dull, unspoken pain when we go *in search of lost instants*. We remember those rich, full hours which ring out with the myriad peals of the Easter bells, the resurrection bells whose sounds we cannot count because in fact all of them count, because each finds its echo in our awakened soul. This memory of joy turns swiftly to remorse when to these hours of total life we compare hours that are intellectually slow because they are relatively poor, hours too that have died because they are empty – empty of all purpose, as Carlyle said from the depths of his sorrow – and hostile hours that, giving nothing, are therefore unending.

And so we dream of that divine hour that would give us everything, not though an *hour that is full* but rather an *hour that is complete*. Here, all the instants of time would be used by matter, all the instants realised in matter would be used by life, and all living instants would be felt and loved and thought. Here, in consequence, the relativity of consciousness would be erased, consciousness being exactly commensurate with complete time.

Finally, *objective time is maximum time*; it is time that contains each and every instant. It is made of all the Creator's many acts, densely entwined.

V

It now remains for us to account for the vectorial character of duration: we need to show what constitutes the direction of time, explaining how it is that a perspective of instants that have gone can be called *past* and how a perspective of something we look towards and expect can be called *future*.

If we have managed to convey the prime meaning of the intuition Roupnel proposes, then it should now be possible to accept that past and future – like duration – correspond to impressions that are essentially secondary and indirect. Past and future do not affect the essence of being and still less the fundamental essence of Time. Let us repeat that for Roupnel, Time is the instant and it is to the present instant that the entire responsibility for time belongs. The past is as empty as the future. The future is as dead as the past. The instant does not hold duration within it, nor does it propel some force in one direction or another. It does not have two faces: it is entire and alone. We can reflect on its essence to our heart's content but we shall never find in the instant the root of a duality which is both sufficient and necessary for us to think a direction.

Moreover when, following Roupnel's lead, we set ourselves to reflect on the Instant, we realise that since we leave one instant only to find another the present does not *pass*. Consciousness is consciousness of the instant and consciousness of the instant is consciousness: these expressions are so similar that they place us in the closest of converses, making it plain that pure consciousness and temporal reality can be assimilated. Once this is grasped in solitary reflection, consciousness has all the immobility of an isolated instant.

It is in the isolation of the instant that time can have poor but pure homogeneity. What is more, this homogeneity of the instant by no means disproves the anisotropy[17] that results from the groupings through which the individuality of durations can be found, the individuality so emphasised by Bergson. In other words, there is nothing in the instant itself that permits us to postulate duration and nothing either that can give us an immediate explanation of our experience – which is real, however – of what we call past and future. For these reasons, we are indeed obliged to try to construct the perspective of instants which alone represents past and future.

Now, when we listen to the symphony of instants, we are aware of dying phrases, phrases that fall away and are borne off towards the past. Yet because of the fact that it is secondary, this flight towards the past is entirely relative. A rhythm only fades away in relation to another section of the symphony which goes on. The following schema is a fairly accurate representation of this relative falling away:

```
 . . .       . .       .
 .....     .....     .....     .....
```

The ratio of three to five *becomes* two to five, then one to five, then silence, the silence of a being that leaves us when all around the world reverberates still.

Using this schema, we can understand all that is both potential and relative in what we call, though without defining its boundaries, the present time. A rhythm that goes on unchanged is a present that has duration; this enduring present is made up of a multiplicity of instants which, from one particular point of view, can be assured of perfect monotony. These kinds of monotony are the stuff of all the enduring feelings that determine a particular soul's individuality. Moreover, unification may take place in the midst of highly diverse circumstances. For the person who goes on loving, a lost love is both present and past, present for the faithful heart and past for the heart in its unhappiness. Such a love is therefore both painful and comforting to the heart that accepts pain and memory together. We are expressing the same thing when we say that a permanent love, that sign of an enduring soul, is something other than joy and suffering and that in transcending emotional contradiction, a feeling that endures acquires metaphysical meaning. A loving soul really does experience the solidarity of regularly recurring instants. And conversely, a uniform rhythm of instants is an *a priori form* of the warmth we feel towards others.

The Instant

The inverse of this first schema would offer us a representation of a rhythm being born and would also give us the elements of a relative measurement of its progress. A musical ear can hear where a melody is going and knows how the new phrase will end. We anticipate the future of a sound, hearing it in advance, just as we do the future of a trajectory, which can be seen in advance. With every fibre in our being, we strain towards the immediate future and it is this straining forward, this tension, that constitutes our present duration. As Guyau has said, it is our intention that in fact orders the future as a perspective in which we are the centre of projection. To quote Guyau, 'we have to desire and we have to want, we have to stretch out our hands and walk in order to create the future. The future is not *that which comes towards us* but rather *that which we go towards*' (*La Genèse de l'idée du temps*, p. 33). Both the meaning and the implications of the future are set down in the present itself.

In this way then we construct in time just as we construct in space. There is a metaphorical persistency here that we shall have to clarify. We shall then recognise that the memory of the past and the anticipation of the future are based on habits. And since the past is only a memory and the future only an anticipation, we shall argue that past and future are both in the end only habits. In addition, these are far from being immediate habits that are established early on. Lastly, like those that make Time appear to us in the perspectives of past and future, the characteristics that make Time seem to us to have duration are not in our view properties seen from the very first. Philosophers have to reconstruct them, basing themselves on the only temporal reality that is immediately given to Thought, that is to say on the reality of the *Instant*.

We shall see that all the difficulties in Roupnel's *Siloë* are centred on this point. These difficulties though may stem from readers' preconceptions. If readers are prepared to get a firm grip on both ends of the chain we are going to put in place, they will arrive later on at a better understanding of how the arguments are linked together. Here then are the two apparently contradictory conclusions we have drawn, and which we shall have to reconcile.

First, duration has no direct force; real time truly exists only in the isolated instant, it is wholly in the actual, in the act, and in the present.

Second, being is however a place where the rhythms of instants re-echo and as such, it could be said to have a past just as an echo has a voice. Yet this past is simply a present habit, this present state of the past being another metaphor. Indeed, habit is not in our view inscribed in matter and space. What we have here is simply a habit that resonates and that remains, we believe, essentially relative. The habit which for us is thought is too ethereal to be recorded and set down, too immaterial also to slumber in matter. It is something that continues, that goes on acting and playing, a musical phrase that has to be taken up again because it is part of a symphony in which it has a role to play. This at least is how we shall attempt to bind past and future closely together by means of habit.

Where the future is concerned, rhythm is of course less firm. There is no symmetry between the nothingness of yesterday and that of tomorrow. The future is simply a prelude, simply a musical phrase that moves on and tries itself out. Just one single phrase. The World is continued by only a very brief period of preparation. In the symphony that is being composed, the future is ensured by just a few bars of music.

Where humans are concerned, the asymmetry of past and future is fundamental. In us, the past is a voice that has found an echo. We thus give force to what is no more than a form, or better, we give one single form to the plurality of forms. And through this synthesis, the past takes on the full weight of reality.

Yet however urgent our desire, the future is a perspective without depth. It has in fact no firm link with reality, which is why we say that the future is with God.

All these ideas will become clearer in a summary of the second theme in Roupnel's philosophy. We refer here to that of habit, which Roupnel's study starts with in fact. Our reason for reversing the order in our own study is this: the absolute negation of the reality of the past is the formidable postulate to be accepted before we can fully appreciate the difficulty of comparing this with common ideas regarding habit. In short, we shall in due course be raising the question as to how the ordinary, everyday psychology of habit can be reconciled with a theory that rejects the direct, immediate action of the past on the present instant.[18]

VI

Before tackling all this however, we could if we wished turn to the realm of modern science for evidence confirming the intuition of discontinuous time. Roupnel did indeed establish a connection between his argument and the description of radiation phenomena in quantum theory.[19] Atomic energy accounts are in fact kept using arithmetic rather than geometry. They are expressed in frequencies rather than durations, the language of '*how often*' gradually replacing that of '*how long*'.

Indeed, at the time Roupnel was writing he could not really foresee the full extent of the theories concerning temporal discontinuity presented at the Solvay Institute's 1927 conference.[20] Reading recent work on atomic statistics, we realise that the question as to which is the fundamental element in these remains unanswered. What should be counted, electrons, quanta, bundles of energy? What is to be regarded as the root of individuality? Going as far back as temporal reality itself in order to find the element mobilised by chance is not an absurd approach. It would then be possible to form a statistical conception of fruitful instants, each instant being taken as isolated and independent from the rest.

There would also be interesting connections to be made between the problem of the atom's positive existence and the way it is always revealed instantaneously. In some respects, radiation phenomena would be fairly satisfactorily explained if we say that the atom exists only in the moment at which it changes. And if we add that this change takes place very suddenly, we are inclined to accept that the whole of reality is condensed in an instant; we should count up its energy using not velocities but impulses.

On the other hand, in showing the importance of the instant in the event, we would be pointing out the considerable weakness of the very frequent objection to the effect that it is the 'interval' separating two instants that has so-called reality. For statistical conceptions of time, the interval between two instants is only an interval of probability; the more prolonged its nothingness, the greater the chance that an instant will come along and put an end to it. It is this intensification of chance that measures its magnitude. Empty duration, duration that is pure, has therefore only chance magnitude. Once an atom no longer radiates, it has an entirely virtual energetic existence: just as it no longer spends anything, the velocity of its electrons consuming no energy, so in this virtual state it does not save power that could be released after a long period of repose. It really is just a toy that has been cast aside when the game is over; even less, it is just a purely formal rule in a game, a rule organising what are simply possibilities. Existence will return to the atom with chance. In other words, the atom will receive the gift of a fruitful instant, receiving it though by chance, as an essential newness, in accordance with the laws of the calculus of probability; this is because the Universe must indeed sooner or later have temporal reality shared out in every part of it, and because too the possible is a temptation to which reality will always finally succumb.

Moreover, chance compels but does not bind with absolute necessity. We can therefore understand how it is that time, which in truth has no real action, gives the illusion of action that must inevitably take place. Were an atom to have remained inactive on numerous occasions while atoms all around it radiated, then it becomes increasingly probable that it is this long slumbering, long isolated atom's turn to act. Repose increases the probability of action but does not really prepare action. Duration does not act 'like a cause',[21] but rather it *acts like a chance*. Here again, *the causality principle is better expressed in the language of the numeration of acts rather than in that of the geometry of actions that have duration*.

All these scientific proofs are beyond the scope of the present study, however. Were we to go into them, we would be taking readers in a direction other than that which is our goal. All we wish to do here in fact is to undertake the task of liberation through intuition. The intuition of continuity often oppresses us and it is no doubt useful therefore to interpret things using the opposite intuition. Whether our proofs are thought to be persuasive or not, the interest of multiplying the different intuitions at the basis of philosophy and science cannot be denied. We ourselves were struck as we read Roupnel's book by the lesson of intuitive independence that was learned in developing a difficult intuition. It is through the dialectic of intuitions that we shall come to use intuitions without running the risk of being blinded by them. The intuition of discontinuous time, taken in its philosophical aspect, helps readers who wish to follow the introduction of theories of discontinuity in highly diverse areas of the physical sciences. Time is what is hardest to think in a discontinuous form. It is therefore by reflecting on this temporal discontinuity realised by the isolated Instant that we shall be given the most direct access to a pedagogy of discontinuity.

Translator's Notes

1 Bachelard's use of epigraphs is a distinctive feature of *L'Intuition de l'instant* (Paris, 1932), marking it out from his previous work and indicative of his long interest in poetry. *L'Intuition de l'instant* is Bachelard's first metaphysical work; published in the same year as *Le Pluralisme cohérent de la chimie moderne* (Paris, 1932) in which he discusses the epistemology of modern chemistry, it follows his first three books, all of which deal with aspects of the philosophy of modern science: *La Connaissance approchée* (Paris, 1928), *Etude sur l'évolution d'un problème de physique: la propagation thermique dans les solides* (Paris, 1928), and *La Valeur inductive de la relativité* (Paris, 1929), the latter being of particular interest here given the influence of relativity theory on his conception of time. Bachelard's reflection on time continues in *La Dialectique de la durée* (Paris, 1936), his second and last metaphysical

work, now available in translation: see *The Dialectic of Duration*, translated and annotated by Mary McAllester Jones, with an introduction by Cristina Chimisso (Manchester, 2000). Six years after *L'Intuition de l'instant*, he turns to write about poetry in *La Psychanalyse du feu* (Paris, 1938), continuing to publish work on science and on poetry until his death in 1962.

The translation given here for Bachelard's first epigraph is that by Keith Bosely, in *Mallarmé, The Poems* (London, 1977); since for the second epigraph, Bachelard gives the author but no source, his version has simply been translated.

2 This book's full title is in fact *L'Intuition de l'instant: essai sur la 'Siloë' de Gaston Roupnel*, and it is intended as a reflection on Gaston Roupnel's *Siloë* (Paris, 1927). In *Siloë* (which has not been translated), Roupnel, a friend and colleague of Bachelard's at the University of Dijon, had taken up the story told in St. John's Gospel (chapter 9, verse 7) of the blind man whose sight was restored after bathing in the pool of Siloam, reinterpreting it in terms of a reflection on human beings, on consciousness, habit, and time, influenced by biology – that of Ernst Haeckel (1834–1919) in particular – by cosmology and by microphysics, all in a spiritualist perspective. Bachelard explains his approach to Roupnel's book in his introduction to *L'Intuition de l'instant*: his aim is not to summarise but rather to experience Roupnel's intuitions, developing them freely in what he terms his own 'arabesques'. Although he does occasionally appear to be influenced by Roupnel's style, Bachelard's ideas are indeed far from his spiritualism: the blind man, in Bachelard's interpretation of the story, symbolises those who refuse to think, who isolate themselves in the Bergsonian consciousness; the restorer of sight, for Bachelard, is reason or more precisely the new reason he perceives in modern science. Jacques Gagey's *Gaston Bachelard ou la conversion à l'imaginaire* (Paris, 1969) discusses in detail Roupnel's effect on Bachelard, his quotations from Roupnel providing a welcome introduction to a work which is not readily attainable (pp. 43–60). Bachelard's thinking in *L'Intuition de l'instant* is discussed and placed in the context of his preceding work on science in Mary McAllester Jones, *Gaston Bachelard, Subversive Humanist. Texts and Readings* (Madison, 1991), pp. 27–38.

3 This would appear to be a rare reference to Bachelard's own life: his wife died in 1920, after a little less than six years of marriage, thirty-eight months of which Bachelard spent in the trenches; after her death, he brought up their daughter alone. Since he makes few personal references of this kind – Bachelard is a discreet writer – this, together with his later reference to lost love, makes the continuing pain of his bereavement very clear.

4 Bachelard is implicitly referring to Bergson's first book *Essai sur les données immédiates de la conscience* (Paris, 1889), translated as *Time and Free Will: An Essay on the Immediate Data of Consciousness*, by F. L. Pogson (London, 1910); Bergson's conception of duration is first presented in the second

chapter of this book and developed the third chapter, in his discussion of freedom and determinism. When Bachelard goes on to refer to life here, he has in mind the general argument of Bergson's *L'Evolution créatrice* (Paris, 1907), translated as *Creative Evolution* (New York, 1998).

5 Bachelard's description of the way the mind works, in particular his reference to refraction and, by implication, to incident rays, emphasises his rejection of the empiricist conception of the mind: when the mind is compared to a mirror, both mirror and mind are simplified; by referring to the physics of mirrors, Bachelard encourages us to think in a new way about the complex relations between mind and reality.

6 Bachelard will present repose differently and more positively in *La Dialectique de la durée* (Paris, 1936), which he in fact describes on the first page as 'an introduction to the teaching of a philosophy of repose'. For details of the translation of this work, see note 1.

7 Bachelard's footnote (amended): cf. Henri Bergson, *Essai sur les données immédiates de la conscience* (Paris, 1889), chapter 2. Bergson writes of 'the gradual enrichment of the self' when discussing duration and the question of its measurability. The word 'fluid' is added by Bachelard.

8 Bachelard's reference is to Bergson's experiment with the sugar-lump in the first chapter of *L'Evolution créatrice* (Paris, 1907). Bergson argues here that the time I wait for the sugar-lump to dissolve is not mathematical time but rather 'it coincides with my impatience, that is to say with a certain portion of my own duration', and is consequently not thought but lived time. At the same time, he seeks to ensure the possibility of objective experience through the coincidence of 'my duration' with that of the universe and also of consciousness with 'organic evolution' by positing at the end of this first chapter a vital impetus in all things, a 'current of consciousness' in all matter, as he puts it in when discussing life and consciousness in the conclusion to chapter 2.

9 Bachelard's footnote (amended): it has already been said by Guyau, though from a more psychological point of view than ours, it is true, that 'the idea of time... is in the end the result of a kind of perspective'; see Jean-Marie Guyau, *La Genèse de l'idée du temps* (Paris, 1890), p. 33.

10 This refers to Maurice Halbwachs's book *Les Cadres sociaux de la mémoire* (Paris, 1925), on which Bachelard also draws in *La Dialectique de la durée*.

11 This may be a puzzling statement for modern readers, accustomed to the power Presidents of France enjoy in the Fifth Republic. It should be remembered that Presidents of France had little power when Bachelard was writing: in 1932 in fact, Albert Lebrun became President, remaining in office until the fall of the Third Republic in June 1940. Alfred Cobban neatly sums up the position of the President in the Third Republic, in particular after Raymond Poincaré's term in office (1913–1920): 'The President's duties were henceforth mainly honorific: he has been described as an

elderly gentleman whose function it was to wear evening dress in day-time'. *A History of Modern France*, volume 3 (London, 1965), p. 21.

12 This conception of time, consciousness, and knowledge can be related to Bachelard's epistemology: see Mary McAllester Jones, *Gaston Bachelard, Subversive Humanist* (Madison, 1991), pp. 27–32; p. 38.

13 The ready use of this mathematical term is indicative of the influence on Bachelard's thinking here of his earlier work on science; an aliquot part is contained an exact number of times in another, dividing it without a remainder, 2 being an aliquot part of 6 for example.

14 The source of this well known quotation is Mallarmé's sonnet 'Le Tombeau d'Edgar Poe'.

15 Louis Couturat (1868–1914), philosopher and logician, is known in particular for his work *De l'Infini mathématique* (Paris, 1896), in which he argued for the actual infinite; this led him to Leibniz, on whom he wrote *La Logique de Leibniz* (Paris, 1901). Couturat played an important part in introducing into France and defending the new mathematical logic of Russell and Peano, especially in his book *L'Algèbre de la logique* (Paris, 1905). Bachelard had referred extensively to Couturat's *De l'Infini mathématique* in his *Essai sur la connaissance approchée* (Paris, 1928), in which he also discussed his theory of fractions

16 Bachelard will return to the notion of a conductor of an orchestra when discussing rhythm, vibration, and matter in the last chapter of *La Dialectique de la durée*, p. 131 (*The Dialectic of Duration*, p. 138).

17 'Anisotropy' is another term indicative of Bachelard's scientific background; anisotropy is defined as the variation of physical properties with direction

18 Bachelard discusses habit in the second chapter of *L'Intuition de l'instant*, entitled 'The problem of habit and discontinuous time'.

19 Bachelard's footnote: cf. *Siloë*, p. 121.

20 Bachelard's reference is to the Fifth Physical Conference of the Solvay Institute, on the topic 'Electrons and Photons', held in October 1927, conferences at the Solvay Institute in Brussels being an important forum for scientific debate. Niels Bohr provides a valuable record of Einstein's contribution to this 1927 conference in a chapter entitled 'Discussion with Einstein on Epistemological Problems in Atomic Physics', in *Albert Einstein: Philosopher-Scientist*, ed. P.A. Schilpp (Evanston, 1949), pp. 200–41. Bachelard also contributed to this volume. In 1927 too, Bachelard presented his doctoral thesis, published in 1928 as *Essai sur la connaissance approchée*.

21 Bachelard's footnote refers to the section in chapter 3 of Bergson's *Essai sur les données immédiates de la conscience*, in which he discusses physical determinism, differentiating between this and 'the realm of life': in the latter, he writes, 'duration does seem to act like a cause'.

5
Time, Instants, Duration and Philosophy

Julian Barbour

This essay is largely a reaction to my first encounter, very incomplete, with the philosophy of time of Henri Bergson and Gaston Bachelard. The editor of this volume, Robin Durie, asked me if I would look at the recently published English edition of Bergson's *Duration and Simultaneity*[1] and Bachelard's essay 'The Instant', which is included in an English translation in this volume.[2] He suggested that I might like to comment on the ideas of these two philosophers and contrast them with my own ideas about time put forward recently in my book *The End of Time*,[3] and also with the ideas in Poincaré's remarkable paper 'The measure of time', published in 1898, and also included in an English translation in this volume.[4]

I start by asking how we can form concepts of time. Both Bergson and Bachelard were keen to identify the true essence of time, and they attempted to do it in the first place from introspection. This seems to me to be a reasonable starting point. In fact, I think it is the only possible starting point if one wishes to theorize about the universe and existence. For we have access to nothing else but the content of experience in consciousness. It seems to me that science and theorizing is possible only because experience is structured. Above all, the basic elements from which the notions of geometry were derived are presented to us directly we open our eyes – we literally see a two-dimensional expanse, and differently colored and textured regions within it, some separated and others contiguous. Our ideas about time are also strongly influenced by our direct experience.

My view about the method of science, or at least theoretical physics, is that we try to explain the phenomena of consciousness by postulating an external world constituted of elements suggested by the content of consciousness but made precise and also 'enlarged'. Let me explain what

I mean by this. Our actual field of vision is, as I said, two-dimensional. However, we feel an ability to move around in the world, which is matched by a continuously changing content of our perceptions. We learn to think in three dimensions. This, instinctively, is a theory of the world, and it involves 'enlargement' from the direct perception of two dimensions to a mental image in three. The fact that our forebears could hunt successfully shows that it is a rather good theory at a qualitative level. It seems that the development of agriculture and commerce was a great stimulus to the development of more precisely mathematical methods: fruit had to be counted, grain and coins weighed, fields measured. This led eventually to the wonderful flowering of Greek mathematics, above all Euclidean geometry. The idea that the universe consists of objects of definite sizes (which may change in time) moving in Euclidean space proved to be a fantastically successful theory.

It is important to understand the method by which theories develop. This was clearly illustrated by the first great success of the scientific method – the theory of planetary motions. The early Greek astronomers were not content merely to record astronomical observations and seek regularities (permitting the prediction of eclipses etc.) in them. They sought to explain their observations. Specifically, they sought to explain the irregular motions observed in two dimensions on the sky by much more regular motions taking place in three dimensions. Definite hypotheses were set up and then tested by observation. In about 150 BC, Hipparchus developed and tested theories of the motions of the sun and moon. They worked quite well. About 300 years later, Ptolemy attempted a theory of the planetary motions using the same geometrical elements and tested his theory against observations. This time, the theory failed. Examining his model and the manner in which it failed, Ptolemy realized that the observations could be explained if he departed from a much-valued principle – that the true motions of the heavenly bodies in three dimensions (as opposed to the manifestly irregular two-dimensional motions directly observed on the sky) are perfectly uniform. He was forced to introduce irregularity, admittedly of a specific nature. His device became known as the 'equant'. For a long time, this was regarded as a defect of his work. Indeed, it was in an attempt to eliminate it that Copernicus hit upon his own scheme, but later Kepler recognized the true worth of the equant, and it played a vital intermediate role in his discovery of the laws of planetary motion.

I mention this episode,[5] to emphasize the essential role of trial and error in the advance of science. Both are essential.[6] Now a trial can only be made with some definite components. Any hypothesis contains various elements. It is here that the theoretician must call upon notions – structural elements – suggested by perception, intuition about the way the world works, and accumulated experience. Then they must be put together to make a definite theory, after which the testing commences. It must also be borne in mind that there is freedom in the choice of what one takes as given and what one is attempting to explain. In the case of time, some of the central issues are the connection between motion (more generally change) and time, the manner in which we are to conceive of instants and of duration, and the connection between them (whether an instant or duration is more fundamental). Because time is so elusive but also all pervading in its effect, it will not be surprising to find very wide differences of approach to the study of time. It is quite illuminating to consider them.

Let us first consider Bergson. The main issue for Bergson seems to be over the status of instants as opposed to duration and the passage of time. However, he is not concerned to show that one is more fundamental than the other within the confines of physics. He has a much more radical position. He argues that the notion of the instant belongs to the domain of quantitative science, while the true essence of time is perpetual flux that can only be grasped by intuition, which he claims is the basis of true philosophy. He asserts that science and philosophy, as he understands it, are quite different approaches to understanding and comprehending the world. And, in his view, the deeper understanding comes from philosophy.

This is not a view to which I should like to subscribe. It seems to me to be basically opposed to the scientific approach. I would certainly not want to claim that the scientific method is the only path to truth, but as of now it is the only path that has good credentials. I believe that many more truths about the world are still to be discovered by the scientific method and that adoption of Bergson's position amounts to acceptance of defeat. I do not see philosophy as a discipline distinct from and superior to science but as the preparatory and initial stages of true science. To it belong the identification of problems that need solution and seem amenable to it by the scientific method, the consideration of alternative strategies, and the search for workable concepts.

To give an example: rumination about the nature of space and geometry – and whether Euclidean geometry is true a priori – was surely part of philosophy for centuries, millennia even. Finally, this inchoate groping, mostly around the axiom of parallels, brought forth true mathematics in the form of non-Euclidean geometries, which was in turn made into brilliantly successful physics by Einstein.

Seen in this light, I think Bergson's *Duration and Simultaneity* does perform a useful role, whereas (despite his immense following, popularity and Nobel prize for literature) his advocacy of pure philosophy of intuition as opposed to mundane science has borne no fruit of which I am aware. Indeed, his faith in it seems to have made him overconfident of the powers of his intuition and led him into an egregious mistake in his debate with Einstein, which I discuss briefly in an appendix to this essay.

The topic on which I want to concentrate for the moment is Bergson's insistence that the scientific notion of the instant fails to capture the most fundamental property of time, its essentially fluid nature. Bergson's claim is that science can at best represent being, not becoming and duration. And for him they are the essence of his *élan vital*, the true force of the universe. In contrast, Bergson felt that an instant is of necessity static. It is like a snapshot. It freezes and thereby kills flux – and with it time. This he felt was intolerable. Since he was convinced that science could only deal with instants, he concluded that the defining aspect of time – flux and becoming – would be forever beyond its purview. Hence his insistence that science could never comprehend time. Even if we do not wish to follow Bergson in his flight to the philosophy of intuition, we can still recognize that he has put his finger on a profound problem. I should like to call it the problem of nouns and verbs, or substance and process. Bergson argued, correctly I feel, that mathematicians instinctively form concepts that correspond to nouns rather than verbs, and make verbs secondary, defining them in terms of change of nouns.

This is illustrated by the definition of a function as a mapping from the elements of one set to the elements of another. Here one could regard the function as representing a process, but this presupposes the elements of the sets, so they – nouns – are primary concepts and the functions secondary. Many distinguished thinkers have shared Bergson's view that process and becoming should be primary, especially his contemporary Alfred North Whitehead, who argued for process in his *Process and Reality*.[7] I have more sympathy with such a viewpoint, since it accepts the scientific method and seeks to make some notion of becoming fundamental. Things are somehow to be derived from processes. This reflects a gut instinct that our sense of change corresponds to something that is more fundamental than that which changes.

I cannot do much more than mention this approach, since, so far as I know, it has never been converted into a scheme that truly works, though

Shimony has made an attempt.[8] Whitehead's book is very difficult. I gave up attempting to understand it. It certainly did not produce anything remotely capable of serving as a scientific model such as Hipparchus provided for the motion of the sun. In the absence of a clear demonstration to the contrary, my own feeling is that we can make a good shot at deriving verbs from nouns but not vice versa. I believe that a scheme, which posits instants as fundamental, does have the potential to explain the appearance of becoming. I should like to give the bare outline of the scheme, which, as we shall see, does in fact involve an element that Bergson insists is central to any notion of time: memory.

Let me start by considering what guidance direct awareness gives us. What hints can we gain from it in forming a concept of time? For myself, by far the most striking fact of which I am aware is the plurality of experience. By this I mean that, in any instant, I am aware of many different things at once. Many different objects appear in my field of vision, and I am absolutely certain that I see them all at once even though only a few are seen really clearly (where my attention is concentrated). In addition, I am aware of sounds, smells, thoughts, and bodily itches that are coincident with my awareness of the visual images. I call such an experience a *plurality within a unity*. Different things seem to be knit together in an indissoluble manner. It is very easy for me to conceive that such a plurality, which in experience is constantly changing, is frozen and becomes a snapshot. I then conceive of change as arising from continual modification of such snapshots. So far as I can see, all the *effects* of change can be more or less perfectly understood in this manner, starting with displacement of the observed position of an object and including any change we can conceive. What cannot be understood and is something genuinely different is our awareness of change as it happens. We literally see motion. We perpetually experience change. The bare notion of an instant, characterized as a snapshot, can give no clue why we experience motion.

Bergson, as noted, takes this as justification for his claim that a scientific account of the world based on the notion of frozen instants will never capture the essence of time and change. Unlike the process philosophers, who strive to find a language of science that does make change primary, Bergson simply walks away from science. In contrast to both these approaches, let me now indicate how far it does seem possible to go in a scheme in which instants are primary.

To start, it will help to say something about space and objects within it. There has been a pronounced tendency, already clearly seen in Greek

Time, Instants, Duration and Philosophy

mathematics, to regard space as uniform extension. It then becomes a container for the objects of the world. Moreover, time is treated as if it were a fourth space dimension. From the mathematical point of view, time becomes identical to space – it is 'spatialized'. Bergson deplores this and says that the true essence of time is thereby lost. However, he does not consider the alternative that I consider more satisfactory. It is to follow the relational view of space and time, most clearly expressed by Leibniz[9] and Mach,[10] as opposed to the absolute view of Newton.[11] To explain the difference, suppose the universe consists of a finite number of particles in Euclidean space. Then one could, with Leibniz and Mach, assert that the instantaneous state of such a universe is exhaustively characterized by the distances between the particles. Newton, in contrast, argued that the totality of the particles also has an overall position and orientation in absolute space. In addition, he argued that such instantaneous states, characterized by the relational distances and the overall absolute data, occur at definite times. This means that identical pairs of states can have very different separations in time.

In the relational view, this is quite wrong. The instantaneous state is simply characterized by the distances between the particles, and time is reduced to change. Two identical relative configurations are simply the same thing, and it is a figment of the imagination to say either that they are differently placed in absolute space or that they occur at different 'times'.[12] However, if they are different, then one can say that their difference is a measure of the 'difference in time' between them. This can, in fact, be translated into a precise mathematical theory of duration. I cannot go into the details, but this is an appropriate place to comment on Poincaré's essay 'The measure of time'. Poincaré notes that there are two fundamental problems in the theory of time. The first was quite widely discussed in the 1890s and concerns the definition of duration: What does it mean to say that a second today is the same as a second tomorrow? However, Poincaré noted that there exists a second problem, barely noted in the 1890s, concerning simultaneity: How is simultaneity to be defined at spatially separated points? This was a startlingly clear anticipation of the problem that Einstein and Poincaré simultaneously solved in 1905 when they created special relativity.

It is unfortunate that the huge excitement generated by the creation of first special and then general relativity led to a concentration of interest on the problem of simultaneity and almost total neglect of the issue of the definition of duration. In this connection, it is especially unfortunate that Einstein himself never thought seriously about the definition of duration. In fact, in my view, Poincaré in essence did outline the best possible definition of duration in his 1898 essay. As I have argued previously,[13] it does, at least in the case of a universe of finite spatial extent, boil down to a measure of the difference of instantaneous configurations of the universe. I think this has an important bearing on Bergson's complaint that the scientific treatment of time reduces it to space. The fact is that an interpretation of duration developed along the lines sketched by Poincaré treats time and space quite differently. Space is expressed through the order that exists within one material configuration of the universe, time is the measure of the difference between configurations.[14]

This leads to an attempt at the representation of reality in which the instant is the most fundamental thing, the ultimate noun. There is no invisible framework or container of the matter in the universe. Instead, there is just a collection of all possible configurations of the universe. Each of them is what, in *The End of Time*, I call an instant, or a *Now*. Nows are clearly 'nouns', not 'verbs'. They are each like frozen snapshots. These relational instants are quite different from the instants that Bachelard, reacting to Bergson, argues should be taken as fundamental. In his essay 'The Instant', Bachelard, having pointed out that relativity 'brought the sudden ruin of ... one unique Duration', makes it clear that he identifies an instant with an event as understood by Einstein: '*the instant, clearly and accurately defined, remains in Einstein's theory an absolute. We can give this absolute value to the instant simply by considering it in its synthetic state, as a point in space-time.*'[15] An event in this sense is a single point in four-dimensional spacetime. Mathematically, it is the exact four-dimensional counterpart of a point in three-dimensional space.

The difference between an event and a Now can be illustrated by imagining a loaf of bread, taken as a model of spacetime. Time runs along the length of the loaf, while space (reduced in this model from three to two dimensions) is in the other two directions. Then an event is just one point in the loaf, while a Now (or spacelike hypersurface) is a complete section through the loaf. It is the entire surface that one would see if the loaf were sliced in half. In Einstein's theory, the three-dimensional Nows each have their own intrinsic geometry. In addition, they are fitted together by virtue of Einstein's equations to make spacetime with a four-dimensional geometry. This is an alternative way to think about Einstein's theory. It was developed quite soon after Einstein's death by Dirac[16] and by Arnowitt, Deser, and Misner (ADM formalism).[17] Wheeler[18] pointed out that this reformulation of Einstein's theory on the basis of three-dimensional spacelike hypersurfaces (my Nows) provides a suitable framework for treating Machian issues. More recently, I have taken Wheeler's suggestion somewhat further.[19] The approach based on Nows leads to different perspectives. In particular, it leads to a satisfactory theory of inertia and duration.

I believe that this understanding of duration (and inertia) does effectively counter Bergson's objection that science reduces time to space. As I said earlier, *space is expressed through the order that exists within one material configuration of the universe, time is the measure of the difference between configurations*. However, this relational resolution would still not satisfy Bergson because it gives no explanation for our firm conviction that time does flow, and, moreover, in a definite direction – into a future and out of a past. We are still stuck with the problem of nouns and verbs – all the most basic precise notions currently employed in theoretical physics correspond to nouns. This is true whether one thinks of four-dimensional spacetime or three-dimensional Nows. It is truly difficult to see how our fluid and often vibrant psychological experience can be recovered from such a stasis. However, rather than adopt a Bergsonian despair of the scientific method, I prefer to await a theory of consciousness and in the meanwhile merely try to get existing science into the best shape possible.

Let me explain what I mean by this. Let us suppose that all conscious experience always has some counterpart in the physical world. This is already accepted, for example, for colours and sounds. They are not constituents of the models that physicists make of the world. But vibrations and frequencies are, and it is such features of the physical models of the world that we have learned to associate with sounds and colours. What feature of the physical models can we associate with our direct experience of change, for example, seeing motion? The most obvious suggestion is that our brains generate the perceived impression of motion by using data on a minimum of two successive positions of the objects that we see moving. In my terminology, the brain creates the impression of motion by comparison of at least two Nows. But a problem now arises. If there is no real flowing time to indicate which Now comes first, how does the brain makes its choice and give us the clear impression that any motion we perceive takes place in one direction and not the other? This is the problem of the *arrow of time*. It was clearly recognized by Boltzmann over a hundred years ago.

His solution was very radical.[20] He recognized that, on purely statistical grounds, complex dynamical systems with many degrees of freedom should with overwhelming probability be in a state of equilibrium, corresponding to a state of maximum entropy and maximum disorder. In such circumstances, nothing distinguishes an 'arrow of time'. Given 'snapshots' of such a system taken successively in 'absolute time', one could imagine them laid out correspondingly along a line. However, a physicist shown the snapshots but not told the direction in which the 'absolute time' increases along the line could not possibly deduce it from the snapshots. Faced with this dilemma, Boltzmann argued that human beings are very complex, highly ordered systems and therefore can, of necessity, only exist when the universe is in an exceptional state of high order and low entropy. According to the laws of dynamics, such states can arise if one waits sufficiently long (very, very long). The system will pass spontaneously from a high-entropy disordered state to a highly ordered state of low entropy after which the order will again decrease and the entropy increase.

But this description has been given under the assumption that there is a direction in which time increases. If this is denied, how could we introduce a 'direction of time' in an unambiguous way? Boltzmann suggested that living beings like ourselves could only do it by means of the degree of order. In accordance with the second law of thermodynamics, the entropy of the world around us is steadily increasing and the order decreasing. He therefore suggested that this is what in fact leads us to say that there is an arrow of time. The 'arrow' that we attribute to an invisible flowing time is actually the arrow of increasing entropy. This has the startling consequence, accepted by Boltzmann, that beings living on opposite sides of the entropy minimum in such a system would point the 'arrow of time' in opposite directions. They would each say that their past lay in the direction to lower entropy and higher order. A direction of time can be unambiguously defined by comparison of the structure in two Nows.

Boltzmann's argument does not cast any real light on the mystery of consciousness and its ability to conjure colours, sounds, and movement out of a monotone, silent and frozen world. But it does suggest to me rather strongly that we can at least hope to make real progress in science by pushing its existing methods just as far as we can. Moreover, it seems to me that to the extent that Bergson's approach appears to have value and explicatory power it is largely because it draws upon the rich possibilities that configurations have for expressing semantic content. I suspect that Bergson is really supping with his devil – who says that everything can be done with frozen snapshots. This is where memory comes into the picture. Consider the implications of this key passage in *Duration and Simultaneity*:

Time, Instants, Duration and Philosophy 105

The mathematician, it is true, will not have to occupy himself with it [i.e., consciousness], since he is concerned with the measurement of things, not their nature. But if he were to wonder what he was measuring, if he were to fix his attention upon time itself, he would necessarily picture succession, and therefore a before and after, and consequently a bridge between the two (otherwise there would be only one of the two, a mere snapshot); but, once again, it is impossible to imagine or conceive a connecting link between the before and after without an element of memory and, consequently, of consciousness.[21]

If we accept that consciousness remains totally baffling, this passage should end at the mention of memory. But then we are extremely close to Boltzmann's position. For how do we understand memory scientifically? Everyone accepts that our memory (certainly our long-term memory) is somehow coded in the trillion or so connections between the several billion or so neurons in our brains. But this is structure in one Now. It is a structure that contains mutually consistent records of what we call our past. Our brain is what, in *The End of Time*, I call a *time capsule*. It is a Now with such a special structure that, by itself, it suggests it is the outcome of a process that has taken place in time in accordance with definite laws. This is how we explain the formation of its records.

When I ponder the implications of brain structure, especially the incredible wealth of information stored in the structure of synaptic connections, it does seem to me that the process philosophers, with their view that verbs are primary, face a very hard task. Their problem, like Bergson's, is the clarity combined with extraordinary flexibility and potential for unlimited complexity that nouns possess. Three black dots on an expanse of white paper are a noun. Being a very simple example, it has little semantic content but is very clearly conceived. In the demesne of such nouns, which populate our imagination at their mere mention, how else can we conceive change but as change of properties, for example as rearrangement of the three dots into a different triangle or as addition or subtraction of a dot? How can we conceive change, transience, flow – in a word, verbs – except as change of nouns? It is the very clarity of the concept of a noun that makes it frozen. Bergson finds this is a bleak prospect, and perhaps it is, but it is compensated by the richness of expressive possibilities. In fact, when Bergson invokes memory, he is actually drawing on them.

Three black dots on white paper convey little. But with hundreds of such dots skilled cartoonists can evoke wonderfully rich situations. Cartoons tell a story. They can express history. They speak of time – and all in a single frozen picture. This is what memory is like. The above extract suggests to me that Bergson failed to understand the significance of instants if properly understood as configurations. The issue is not measurement but meaning. Instants – Nows – are like snapshots, and they can be immensely evocative. When we look at a snapshot of ourselves as a child happy with toys, what a world of play and motion is instantly conjured up. Are we measuring? No, we are interpreting and comprehending. So for me time is something we use in interpreting and ordering Nows. I shall have to refer the reader to *The End of Time* and to 'The Timelessness of Quantum Gravity' for the details. In these comments, I merely wanted to indicate how hard it is to think without nouns.

If the followers of Bergson and Whitehead can ever supplant nouns by verbs, they will have wrought a truly great revolution in thought. I doubt if it can be done.

Appendix: on Bergson and the twin paradox

Bergson's *Duration and Simultaneity* contains a strong attack on the multiple relative times (durations) that appear in Einstein's special relativity. The book, which Bergson would have done well to suppress, reveals a profound misunderstanding of the principle of special relativity. What this actually states is that the complete universe can be described in certain distinguished frames of reference, called inertial frames. These frames are singled out by the fact that in them the laws of nature take on a special simple form. The existence of these frames is the first fundamental fact postulated by the principle. The second is that there is a whole family of exactly equivalent frames of this kind. Given any one such frame, any other frame moving relative to it with uniform velocity in a straight line is also an inertial frame. All the laws of nature take exactly the same form in any inertial frame. Any process observed to unfold in a particular manner in one such frame can be observed to unfold in the same manner in any other such frame. However, they are different processes, unfolding in different parts of the universe.

Let me illustrate this by the notorious twin paradox, which is the particular cause of Bergson's mistake. Consider quadruplets that initially are moving in pairs (Peter and Martha in the first pair, Mary and Paul in the second) relative to each other at about 85% of the speed of light. All four quadruplets carry identical clocks. Let the pairs pass each other at some point in spacetime, when all clocks are set to zero. Each pair is

Time, Instants, Duration and Philosophy

associated with an inertial frame of reference, which is given for all time. To underline this fact, let Peter remain for ever in his inertial frame and Mary remain for ever in her frame. Only two frames of reference are used in this story. At the given relative speed, the equations of special relativity tell us that in Peter's frame the clocks in Mary's frame tick at only half the rate of Peter's clocks. But equally in Mary's frame the clocks in Peter's frame tick at only half the rate of the Mary clocks. This seeming impossibility is made possible by the fact that Mary's axes are laid out skew over the universe as described in Peter's frame and vice versa.

After the Mary clocks have ticked 50 times, Paul, who (with Mary) was rushing away from Peter and Martha at 85% of the speed of light, changes speed instantaneously and now moves toward Peter at 85% of the speed of light. His clock will again run at only half the rate of the Peter clocks. So when he meets Peter, his clock will have ticked 100 times, while Peter will find his clock has ticked 200 times. This is the twin paradox – Peter has aged twice as much as Paul. People have difficulty with this result for two quite different reasons. There are those who are simply convinced that time flows at the same rate for all real physical objects (including living beings) irrespective of their motion in the universe. At least as regards the actual physical manifestations of time (as opposed to the psychological experience of the flow of time), Bergson adopts this position in *Duration and Simultaneity*. Then there are others, including Bergson, who find that the result seems to confound the basic premise of special relativity – that all statements that can be made in one inertial frame of reference can be equally made in another. From this they conclude that since Peter and Paul always have the same relative velocity, each must always be able to say that the clocks of the other are going slower than their own. When they meet, each should be younger than the other. This is clearly impossible, hence the 'paradox'.

The mistake that these people make is that they misunderstand special relativity. The inertial frames of reference are given once and for all. Paul moves from one to another, and this establishes an ineradicable asymmetry between Peter and Paul. It can never be restored for the unfolding of the Peter—Paul story. The story needs to be told consistently in the Peter frame. This is what Bergson obstinately refused to see. He repeats over and over again that, precisely by the rules of relativity, either Peter or Paul can regard themselves forever as at rest, defining thereby equally valid frames of reference. This is simply to misunderstand relativity. Such symmetry does not exist. However, if symmetry can never be established between Peter and Paul, it is possible to find another story that exactly matches the Peter—Paul story and does restore symmetry.

This is achieved as follows. After the clocks of Peter and Martha have ticked 50 times, Martha sets off to catch up with Mary and Paul, who are receding from her at 85% of the speed of light. Martha chooses her velocity, which must be faster than this, so that when she does catch up with Mary (who carries on in the Mary frame after Paul has left her to join up with Peter) Mary's clock will have ticked in total 200 times. In fact, a simple calculation shows that Martha's speed is so great in the Peter frame that it ticks in it about seven times more slowly than the Peter clocks. In the Mary frame, it ticks half as fast as the Mary clocks. This ensures that when Martha and Mary meet, Martha's clock has ticked 100 times, Mary's 200 times. As regards the observable outcomes (comparison of clocks on meeting), there is now perfect symmetry between the Peter (200 ticks)—Paul (100 ticks) story and the Mary (200 ticks)—Martha (100 ticks) story. In each frame, there are two different stories that unfold over different parts of spacetime. In the Peter frame, the Mary—Martha story ends much later than the Peter—Paul story, but in the Mary frame it is the other way round. Thus, if one considers the pair of stories as one single process, identical processes unfold in each frame. In addition, it is always the clock that travels on the straight line in either story that ticks more – and its carrier ages more.

Failing to appreciate that the symmetry of relativity applies to a pair of such stories, Bergson tries desperately to find symmetry in just one. The only way he can do this is to claim, in complete misunderstanding of the theory, that in the Peter frame only one clock is 'real'. This is the clock that travels with Peter, who is a real living physicist whose consciousness actually registers the ticks of his clock and ages in step with it. Bergson claims that all the other clocks and ticks are nominal, merely things that Peter calculates for an imaginary Paul in order to give a pleasing symmetric description of the world. They are not real things that real conscious people would experience. Reading the main text of *Duration and Simultaneity*, where such statements are made repeatedly, one wonders how he will deal with the twin paradox. That only comes in the Appendices added later. In them, Bergson does truly stick his neck out – too far for his reputation unfortunately.

In Appendix 1, he says:

> all this applies not to a living, conscious Paul, but to a Paul whom physicist Peter *pictures as watching* this clock ... It is for this merely imagined and

referred-to Paul that four – imagined – hours will have elapsed while eight – lived – hours will have elapsed for Peter. But Paul, conscious and therefore referrer, will have lived eight hours, since we shall have to apply to him everything we just said about Peter.

Now the twin paradox experiment with real living people has not yet been tested. The effects at currently attainable speeds are still too small to detect in living organisms. Thus, were Bergson still alive, he might just argue that his claim has not yet been disproved. However, in Appendix III, Bergson made a claim that has now been unambiguously refuted – that what he claimed above for the living Peter and Paul would also hold for inanimate clocks. This is what he says:

> In reality we should say that the moving clock exhibits this slowing at the precise instant at which it touches, still moving, the motionless system and *is about to re-enter it*. But, immediately upon re-entering it, it points to the same time as the other ... For the slowed time of the moving system is only attributed time; this merely attributed time is the time indicated by a clock hand moving before the gaze of a merely imagined physicist ... [it is] a phantasmal clock, substituted for the real clock throughout its journey: from phantasmal it again turns into real the moment it is returned to the motionless system. It would, moreover, have remained real for a real observer during the trip. It would not have undergone any slowing. And that is precisely why it shows no slowing when it is again found to be a real clock upon arrival.

Unfortunately for Bergson, just such an experiment was already performed decades ago with real atomic clocks on real passenger aircraft flown around the world. The outcome supported Einstein, not Bergson.

College Farm,
South Newington, Banbury,
Oxon, OX15 4JG, UK

Notes

1 Henri Bergson, *Duration and Simultaneity: Bergson and the Einsteinian Universe* (trans. L. Jacobson, M. Lewis & R. Durie, ed. R. Durie; Manchester: Clinamen Press, 1999).
2 Gaston Bachelard, 'The Instant' (trans. M. MacAllester Jones); see Chapter 4 of this volume.
3 Julian Barbour, *The End of Time* (London: Weidenfeld & Nicolson, 1999).
4 Henri Poincaré, 'The Measure of Time' (trans. G.B. Halsted); see Chapter 2 of this volume.

5 Full details of which can be found in Julian Barbour, *Absolute or Relative Motion?* Volume 1, *The Discovery of Dynamics* (Cambridge: Cambridge University Press, 1989). A paperback edition is to be published by Oxford University Press, in 2001.
6 The best summary of the scientific method that I know is the one given by David Deutsch in *The Fabric of Reality* (London: Penguin Press, 1997).
7 A.N. Whitehead, *Process and Reality* (Corrected edition edited by D. R. Griffin & D. W. Sherbourne; New York: The Free Press, 1968).
8 Abner Shimony, 'Implications of transience for spacetime structure', in: *The Geometric Universe: Science, Geometry, and the Work of Roger Penrose* (edited by S. A. Huggett et al.; Oxford: Oxford University Press, 1997).
9 See for example *The Leibniz—Clarke Correspondence* (edited by H. G. Alexander; Manchester: Manchester University Press, 1956).
10 Ernst Mach, *Die Mechanik in ihrer Entwicklung. Historisch-kritisch dargestellt* (J. A. Barth, Leipzig, 1883); English translation: *The Science of Mechanics: A Critical and Historical Account of its Development* (LaSalle: Open Court, 1960).
11 Isaac Newton, 'Scholium on absolute space and time', in: *Philosophiae Naturalis Principia Mathematica* (Royal Society, London, 1687); English translation: *Newton's Principia* (Berkeley and Los Angeles: University of California Press, 1962).
12 See for instance §§ 5ff of Leibniz's Fourth Letter to Clarke, and §§ 4–6 of his Third Letter, in *The Leibniz—Clarke Correspondence*.
13 Julian Barbour, 'The timelessness of quantum gravity: I. The evidence from the classical theory; II. The appearance of dynamics in static configurations', *Classical and Quantum Gravity* 11 (1994), 2853. See also Chapter 6 and pp. 135–6 of *The End of Time*.
14 See Barbour, 'The timelessness of quantum gravity: I. The evidence from the classical theory; II. The appearance of dynamics in static configurations'.
15 Bachelard, 'The Instant', § III (p. 75 of this volume) [*L'Intuition de l'instant* (Paris: Editions Stock, 1931), pp. 30–1]
16 P.A.M. Dirac, 'The theory of gravitation in Hamiltonian form', *Proceedings of the Royal Society (London)*, A246 (1958), 333.
17 R. Arnowitt, S. Deser, & C.W. Misner, 'The dynamics of general relativity', in: *Gravitation: An Introduction to Current Research* (edited by L. Witten; New York: Wiley, 1962).
18 J.A. Wheeler, 'Mach's principle as boundary condition for Einstein's equations', in: *Gravitation and Relativity* (edited by H.-Y. Chiu & W.F. Hofmann, New York: Benjamin, 1964); 'Geometrodynamics and the issue of the final state', in: *Relativity, Groups, and Topology: 1963 Les Houches Lectures* (edited by C. DeWitt and B. DeWitt; New York: Gordon & Breach, 1964).
19 See Barbour, *The End of Time* and 'The timelessness of quantum gravity: I.

Time, Instants, Duration and Philosophy 111

The evidence from the classical theory; II. The appearance of dynamics in static configurations'.
20 Ludwig Boltzmann, 'On certain questions of the theory of gases', *Nature* 51, (1895), 413; "Zu Hrn. Zermelos Abhandlung 'Über die mechanische Erklärung irreversibler Vorgänge', *Wied. Annalen* 60 (1987), 583.
21 Bergson, *Duration and Simultaneity*, p. 33 [Bergson, *Mélanges*, (ed. A. Robinet; Paris: Presses Universitaires de France, 1972) p. 101].

6
The Present Moment in Quantum Cosmology: Challenges to the arguments for the elimination of time[1]

Lee Smolin

Abstract

Barbour, Hawking, Misner and others have argued that time cannot play an essential role in the formulation of a quantum theory of cosmology. Here we present three challenges to their arguments, taken from works and remarks by Kauffman, Markopoulou and Newman. These have a common element, which concerns the realization by finite procedures of formal constructions used in the arguments for the elimination of time. This leads also to a new solution to the problem of time, based on a new kind of dynamical theory in which the construction of the configuration space cannot be carried out in advance of the evolution of the solutions of the theory.

1. Introduction

There is a general agreement that the notion of time is problematic in cosmological theories. During the last years discussions about time in cosmology have tended to focus on the question of whether time can be eliminated from the fundamental statement of the laws and principles of the theory. The argument that time can be eliminated has been put forward in its strongest form to date by Julian Barbour in *The End of Time*.[2] I am convinced by Julian's argument. If a quantum cosmological theory can be formulated along the lines contemplated by many people in the field it follows, as he has argued, that time can be eliminated from the theory. One of my first tasks here will be to outline the logic of the argument and show that it does indeed follow from the premises usually assumed when talking about quantum cosmological theories.

However, before accepting the conclusion we should ask whether the premises are correct or not. The main part of my argument will be that

The Present Moment in Quantum Cosmology 113

there are strong reasons to doubt the correctness of several of the key premises. These come from the unjustified reliance of the proposed quantum theories of cosmology on mathematical structures that have no relevance for the representation of observations that can be made by real observers inside the universe. Once these are eliminated it is no longer possible to make Barbour's argument. However it also becomes necessary to find new ways to formulate a fundamental theory of cosmology that does not run afoul of the problems.

Thus, the aim of my essay is to argue that the problem of time is genuine, but that its resolution requires, not the elimination of time as a fundamental concept, but instead a reformulation of the basic mathematical framework we use for classical and quantum theories of the universe. This reformulation is in any case necessary to produce a theory in which every quantity that is formally observable is in fact something that could be measured by observers inside the universe. The end result, I will argue, is that time will play an even more central role in the formulation of fundamental physics.

My argument will have four parts. In the first, I review the basic structure of classical and quantum theories of cosmology and show that such theories are characterized by five postulates. In the second I sketch the basic argument why time is not a fundamental concept in theories which satisfy the five postulates. I believe that this account strengthens Barbour's claim that time is not a fundamental concept in theories that satisfy these postulates. In the third part, I will give three objections to the argument, which attack one or more of the postulates. In the final part I sketch the framework of a dynamical theory which takes these objections into account and could serve as a quantum cosmological theory.

In order to make the structure of the argument clear what follows is only a quick sketch of the argument. There is much more that could be said on the subject. However my view is that this question is not one that can be settled by philosophical argument alone. Were that possible the problem would already have been resolved. The point of my argument is that the problem of time both requires and points to a change in the structure of our physical theories. What needs to be done next is to see if theories of this kind can be constructed and if they may also help to resolve other issues in physics such as quantum gravity and the foundations of quantum theory. I believe that the answer is likely yes, but this is something that cannot be argued for, it must be tried to see if it succeeds or fails.

Very little of my argument is original. Julian Barbour has greatly strengthened the argument for the elimination of time in cosmological theories, and the argument here was developed mainly in reaction to his recent papers.[3] I disagree with his conclusions but see no escape short of the kind contemplated here.

The arguments against the elimination of time I present here are due to Stuart Kauffman,[4] Fotini Markopoulou[5] and Ted Newman.[6] Discussions with a number of other people, especially Jeremy Butterfield, Saint Clair Cemin, Louis Crane, Fay Dowker, Chris Isham, Louis Kauffman, Jaron Lanier, Seth Major, Carlo Rovelli, Rafael Sorkin and Simon Saunders have been crucial for my understanding of these issues.

Finally I must stress that both the positive and negative steps of my argument do not depend very much on the details of the cosmological theory under consideration. They apply as strongly in string theory as in quantum general relativity and they appear in all versions of these theories so far known.

1.1 Some questions

I would like to preface my argument with several basic questions about cosmology:

- What is an observable in a cosmological theory, based on general relativity? What is the actual physical content of the theory and how can we separate it out from its mathematical presentation?
- What will be an observable in the ultimate theory of quantum cosmology?
- Is novelty possible? Does general relativity allow the existence of entities that do not exist at some time, but exist at some other time (where a moment of time is identified with a spacelike surface)?
- Will the ultimate quantum theory of cosmology allow the possibility of novelty?

The importance of the first question is that presently we can give a formal characterization of observables, but we are actually not able to construct many examples of quantities that satisfy it. This stems from the fact that general relativity does not, as is often said, identify the history of a physical universe with a manifold on which are defined a metric and perhaps other fields. The correct statement is that the history of a universe is defined by *an equivalence class of manifolds and metrics under arbitrary diffeomorphisms.*[7]

This is a key point; the significance of it is still often overlooked, in spite of the fact that it is far from new. One major consequence of it is that there are no points in a physical spacetime. A point is not a

The Present Moment in Quantum Cosmology

diffeomorphism invariant entity, for diffeomorphisms move the points around. There are hence no observables of the form of the value of some field at some point, x.

If observables do not refer to fields measured at points what in the world do they refer to? We have to begin with only the characterization that an observable must be a functional of the fields on a manifold, including the metric, which is invariant under the action of arbitrary diffeomorphisms.

This is easy enough to state. It is harder, but still possible, to describe a few observables in words. For example the spacetime volume is an observable for compact universes. If the theory contains enough matter fields one can attempt to use the values of some of the matter fields as coordinates to locate points in generic solutions. Once points are labelled by fields, the argument goes, they have a physical meaning and one can then ask for the values of other fields at those points.

There are however several unsolved problems that make it doubtful that this is a satisfactory way to describe the observables of the theory. Past the first few simple ones such as spacetime volume we do not know whether the others are actually well defined on the whole space of solutions to Einstein's theory. For example all attempts made so far to use the values of some physical fields as coordinates on the space of solutions fail because of the unruly behaviour of the fields in generic solutions. As a result, we have control over only a handful of observables. But we need an infinite number of observables if we are to use them to distinguish and label the infinite number of solutions of the theory.

Finally, we must ask if any of the observables are actually measurable by observers who live inside the universe. If they are not then we cannot use the theory to actually explain or predict any feature of our universe that we may observe. If we cannot formulate a cosmological theory in terms that allow us to confront the theory with things we observe we are not doing science, we are just playing a kind of theological game and pretending that it is science. And the worrying fact is that none of the quantities which we have control over, as formal observables, are in fact measurable by us. We certainly have no way to measure the total spacetime volume of the universe or the spacetime average of some field.

The reader may ask what relativists have been doing all these years, if we have no actual observables. The answer is that most of what we know about general relativity comes from studying special solutions which have large symmetry groups. In these cases coordinates and observables can be defined using special tricks that depend on the symmetry. These methods are not applicable to generic spacetimes; furthermore there is good reason to believe that many observables which are defined for generic spacetimes will break down at symmetric solutions.

Thus, relativists have sidestepped the problem of defining observables for general relativity and solved instead a much simpler problem, which is defining observables for fields moving on certain fixed backgrounds of high symmetry. It is fair to say that the result is that we do not really understand the physical content of general relativity; what we understand is instead the physical content of a series of related theories in which fields and particles move on fixed backgrounds which are themselves very special solutions of Einstein's equations.

The questions about novelty get their relevance from the fact that we observers in the real universe do genuinely observe novelty, in the sense that we observe that things happen which could not have been predicted on the basis of all the information that was, even in principle, available to us. One source of novelty is that each year new stars and galaxies come into view from which we could not, on the standard models of cosmology, have received light signals, due to the universe's finite age and the finiteness of the speed of light. One may try to evade this argument in the context of an inflationary model, but this requires that we be able to predict the precise details of the light received from these distant galaxies on the basis only of the physics of their past during the inflationary era. But this is impossible in principle as the patterns of inhomogeneities that, according to the models of inflation, become the seeds for galaxy formation, are themselves seeded by quantum fluctuations in the vacuum state of a field during the inflationary era. Thus it is impossible in principle to predict the light that we will see next year arriving from a star presently too far to see, even assuming that inflation is correct.

A second source of novelty has to do with the fact that we live in a complex universe, so that we are constantly confronted with novel biological, sociological and cultural phenomena. It may be obvious that this can have nothing to do with the problem of time in physics, but there are some related issues that we will need to consider carefully later in this essay.

In any case the first source of novelty is genuine and this is worrying enough. How are we to reconcile the fact that there is a necessarily unpredictable component to what we observe with the claim that time can be eliminated from our fundamental cosmological theories? There may be an answer to this question, but this is an issue we will have to look at closely before judging the claim that time can be eliminated from physics.

2. The arguments for the elimination of time in cosmological theories

Before giving the arguments for the elimination of time I must emphasize a crucial point, which is that they concern cosmological theories. There is no problem of time in theories of isolated systems, embedded inside the universe or theories of systems with boundaries, at either finite or infinite distances. The reason is that if the system modelled is understood to include only part of the universe one has the possibility of referring to a clock in the part of the universe outside the system which is modelled by the theory. This is generally what the t in the equations of classical and quantum mechanics refers to.[8] The problem of time arises only in cosmological theories in which the whole universe is included in the degrees of freedom modelled in the theory, so that any clock, and any measuring instruments referred to in the interpretation of the theory must be part of the dynamical system which is modelled.

For the purpose of this discussion a universe is a closed system, which contains all that any part of it may interact with, including any observers and observing instruments and any clocks used to measure time. We will call any description of the physics of such a system a cosmological theory whether or not it is believed to be the actual universe, as its description faces the formal problems we are concerned with here.

There are two closely related arguments for the elimination of time in cosmological theories, the first classical, the second its consequence for quantum theories of cosmology.

2.1 The standard framework for classical cosmological theories

The argument for the elimination of time in classical physics begins with the definition of a *configuration* of a universe. A configuration is a possible state or situation that the universe can have at a given moment of time. The arena in which the argument takes place is the *configuration space* of the universe, denoted C. This is defined to be the space of all possible *configurations* that the universe can be in at a given moment of time.

For a single particle restricted to be in a room the configuration space, C_{room} is defined to be the three-dimensional space contained by the room; each point in the room is a place the particle might be and hence a possible configuration of the system. For a system of N particles in the room the configuration space is C_{room}^N.

In classical physics it is assumed that universes have *histories*, which are described as curves $x^a(s) : R \leftarrow C$ in the configuration space. $s \in R$ is called the *time parameter*. Part of the question we are concerned with is the role of these histories in the description of the universe and the relationship between time parameters and time as measured by a clock carried by an observer inside the universe. A closely related question is the relationship between the time parameters and our experience of the flow of time. For the purposes of this paper we will be able to ignore the second question, as the answers to the first question that will be considered will change considerably the context for consideration of the second.

The application of classical physics to cosmology is based on the following four postulates:

- **Postulate A: Constructibility of the configuration space.** It is possible by a finite mathematical procedure for observers inside a universe to construct its configuration space C.
- **Postulate B: Deterministic evolution.** There is a law for the evolution of universes which, given a position $x^a(0) = p^a \in C$ and a velocity vector, $v^a \in T_p$ at p^a, picks a trajectory $x^a(s)$ which is unique up to redefinitions of the time parameter S. The law is given by an action principle $S[x^a(s), v^a(s)]$.
- **Postulate C: Observability of the configuration space.** It is possible for observers inside a universe to make measurements which are sufficient to determine which trajectory $x^a(s)$ describes the history of our universe.

We may note that each postulate is necessary for cosmology to be treated according to the usual methods of classical physics. If **A** is not satisfied then we observers inside a universe cannot describe the universe according to the methods of classical physics. If **B** is not satisfied then we cannot use the theory to make predictions of the future or retrodictions of the past. If **C** is not satisfied then we cannot compare those predictions and retrodictions with observations.

Given the definition we have specified of a universe as a closed system, which contains all that any part of it interacts with, there is no role whatsoever in the notion of *an observer outside the universe*. Any recourse to such an observer in discussions of cosmological theories represents an implicit admission that the theory under discussion is not a proper framework for doing cosmology. For this reason we take over the principle enunciated by Markopoulou:[9] *All observables in a cosmological theory must be measurable by some observer inside the universe.*

The Present Moment in Quantum Cosmology

All modern approaches to cosmology take on board the principle that observations of the configuration of the universe are relational in the sense that they refer to coincidences in the values of variables observable from inside the universe. This principle was first enunciated by Leibniz,[10] and the exact sense in which it is satisfied by general relativity is explained by Barbour in *The End of Time*. We may express this as a fourth postulate:

- **Postulate D:** C **is a relative configuration space.** This means that a point $X^\alpha \in C$ is completely determinable by measurements made by observers inside the universe.

This has the important consequence that, as any two points x^a and y^a of C must refer to physically distinct configurations that can, moreover, be distinguished by observers inside the universe, there can be on C no global symmetry such as Euclidean invariance and no local gauge invariance such as diffeomorphism invariance.

The exact form of the relative configuration space depends on the content of the theory. For a system of N particles in d dimensional space, the relative configuration space may be constructed as a quotient,

$$C_{rel} = \frac{R^{dN}}{\mathbf{E}} \tag{1}$$

where $\mathbf{E}$ is the Euclidean group of rotations and translations in R^d.[11]
For general relativity and related theories such as supergravity,

$$C_{gr} = \frac{\text{metrics and fields on } \Sigma}{Diff(\Sigma)} \tag{2}$$

where $Diff(\Sigma)$ is the group of diffeomorphisms on a compact manifold Σ, which is taken to represent 'a spatial slice of the universe.'

An important point, to which we will return later, is that the relative configuration spaces are, at least in the examples studied so far, defined as quotients of a well defined space by the action of a group.

Nor can there be any notion of a clock which is something other than a degree of freedom measurable by observers inside the universe. Since all observables are assumed to be relational, and hence to measure coincidences in values of measurable quantities, there can be no reason why one clock can be preferred over another, so long as the values of the two of them can be related to each other uniquely. The result is our final postulate.

- **Postulate E: Reparameterization invariance.** Two trajectories $x^a(s)$ and $x^a(s')$ which differ only by a redefinition of the time parameter, $s' = f(s)$ are deemed to describe the same physical history of the universe. This implies that the action $S[x^a(s), v^a(s)]$ is invariant under these redefinitions.

2.2 The classical argument for the elimination of time

The five axioms we have given define a classical cosmological theory. They rule out the existence or relevance of any clocks outside the system. Postulate **E** also rules out any absolute internal time, in that all time parameters appear, at least at first, to be on an equal footing. We must then wonder what the proper notion of time is in such a theory and, in particular, if it contains any concept of time that can be connected with our observations.

I will claim, following arguments by Barbour and others, that the theory allows no fundamental notion of time. By this we mean that the theory can be formulated in such a way that no reference is made to a time parameter. This does not exclude the introduction of parameters which have, at least in particular solutions, some of the properties of time in ordinary theories. Often these are degrees of freedom which behave in some regimes like physical systems we call clocks. We may thus call these clock variables. Typically these behave as would be expected in some regions of the space of solutions, but not in all. For these reasons, they may provide an approximate or effective notion of time in some domains in some solutions. But if the theory can be formulated without any notion of time then there is no guarantee that these approximate clock variables measure some more fundamental quantity which deserves the name of a universal time for the theory as a whole.

The basic argument for the elimination of time is the following. Given **B**, any pair $(x(s), v^a(s))$ determines a unique history $x^a(s)$. Conversely, given any history determined by the laws, and an arbitrary choice of initial time labelled by $s = s_0$, the whole history of the universe $x^a(s)$ is determined by its initial data $(x(s_0), v^a(s_0))$. But any observable O measurable by an observer inside the universe must be a function of the pair $(x(s), v^a(s))$. As a result any such function is actually determined completely by the initial data $(x(s_0), v^a(s_0))$. This means that

The Present Moment in Quantum Cosmology

any observable must have the same value on any two points of the same trajectory.

This conclusion may seem counterintuitive when first encountered, and so it may help to go over the argument a bit more carefully. The point is that the time parameter s is not measurable by any observer. Because it can be changed without consequence to the physical history, its only role is as a mathematical device to label the different points on a trajectory. Since that labelling is arbitrary, it cannot correspond to anything that an observer inside the universe can measure; in particular it cannot correspond to the reading on a physical clock.

Something which is observable then must be expressed as a correlation between different functions of the pairs $(x^\alpha(s), v^a(s))$ that is independent of the parameterization s. For example it may be a correlation between the reading of one dial of an instrument and another. These correlations are independent of the parameterization and can be used to define a physical notion of time which is observable by an observer inside the system. But as it is independent of the parameterization, such a correlation must be determined by the $(x^\alpha(s_o), v^a(s_o))$ for any arbitrarily chosen parameter value s_o. But since s_o is arbitrary this means the observable is actually a function of the trajectory. If it is to be expressed in terms of the pair $(x(s), v^a(s))$ it must be in a way that is constant along each trajectory.[12]

Mathematically this has the following consequence. Because each pair $(x(s), v^a(s))$ determines a unique history, the evolution along each trajectory defines one parameter group of diffeomorphisms E of C, each member of which takes each point in C to another point on the same history. The argument just given says that any physical observable O must be invariant under the action of E. As a consequence one can take the quotient

$$U = \frac{C}{E} \qquad (3)$$

to define H, the space of histories. But it then follows that any observable O is actually a function on U.

The final form of the theory is then that the possible universes are described as points of U. All observables are functions on U. The notions of time, trajectory and history have disappeared from the final form of the theory.

The definition of the quotient may involve some technical work, but is assumed to be always possible. One way to do it is to define a gauge condition, which is an equation for a surface in C that intersects each history exactly once. One may then identify U with this surface. Physically, one way to do this is to define a physical time parameter in terms of some observable quantity $T(x^a)$ which is a function on C which has the property that $(x^a) = 0$ defines such a surface.

The result is that time has been eliminated completely from the theory. One does physics by determining which trajectory one is on and then determining the value of the observables, expressed as correlations between quantities measurable by local observables. Some of these quantities may be interpretable as readings on devices we call clocks, in which case we can recover at some level a notion of time, defined just as the readings on a device called a clock. But there is no reason we must interpret the observables in terms of such clock variables. Time may or may not be a useful construct, good for some level of description. But it has no fundamental role in the theory.

2.3 The argument for the elimination of time in quantum cosmology

We now turn to discussion of the quantum theory. We will take the conventional view that a quantum theory of cosmology should be a quantization of a classical theory of cosmology. This is of course unlikely to be true, as the quantum theory is assumed to be the fundamental theory and the classical theory should be derived from it by some suitable approximation procedure. But we will adopt the conventional view here, as it is the context in which the argument for the elimination of time in quantum cosmology follows. As before I will only sketch the argument.[13]

There are several different approaches to turning a classical cosmological theory into a quantum theory. We describe only one here, as the logic in the other approaches is similar. This is the approach based on the Wheeler-DeWitt, or *hamiltonian constraint equation*. This approach arises in the context of hamiltonian quantization of theories which satisfy postulate **E**, invariance under arbitrary redefinitions of the time parameter. To make the argument as transparent as possible to those unfamiliar with it, and to avoid retreading familiar ground for those who are, I skip the steps of the construction and exhibit only the result.

In this approach, a quantum theory of cosmology depends also on five postulates. Postulate **A** is taken over directly from the postulates of classical cosmology. The postulates **B** and **C** are replaced by

The Present Moment in Quantum Cosmology

- **Postulate B′: Existence of the wavefunctional of the universe.** A quantum state of the universe is defined to be a normalizable complex functional Ψ on C. This means that there exists a measure $d\mu$ on C such that

$$1 = \int_c d\mu \ |\Psi|^2 \tag{4}$$

The normalizable states define a Hilbert space on which μ defines an inner product according to

$$<\Phi | \Psi> = \int_c d\mu \ \Phi^*\Psi \tag{5}$$

Observables must then be represented as self-adjoint operators in this inner product space. As in ordinary quantum mechanics there must be complete commuting sets of observables whose spectrum determine the quantum state uniquely. This gives rise to

- **Postulate C′: Observability of the configuration space.** It is possible for observers inside a universe to make measurements which are sufficient to determine which wave-function Ψ describes the history of our universe.

Postulate **D** is either taken over directly, or replaced by a set of first order functional differential equations called the diffeomorphism constraints $D(v)$, which depend on an arbitrary vector field v^a on Σ, the spatial manifold. This expresses the requirement that the quantum states are valued only on relative configurations which are in this case diffeomorphism equivalence classes of metrics on Σ. In the latter case we have:

- **Postulate D′: Diffeomorphism constraints.** The wavefunctionals satisfy also

$$D(v) \ \Psi = 0 \tag{6}$$

for all vector fields v^a on Σ.

Finally, the crucial postulate **E** that expresses time reparameterization invariance is replaced by:

- **Postulate E′: Solution to the Hamiltonian constraint.** There is a second order functional differential operator, H on C such that

$$H \Psi = 0 \tag{7}$$

Of course many technical problems arise in the realization of these conditions. An important motive for taking them seriously is that they can in fact be realized in quantum general relativity in 3+1 dimensions, coupled to arbitrary matter fields.[14]

There are many interesting issues concerning the construction and interpretation of such observables. We do not need to go further into this discussion than to note that as in the classical theory time plays no role in the actual formulation of the theory. As many have remarked, the Hamiltonian constraint equation, (7) contains the dynamics of the theory, but there is no d/dt on its right hand side. This is because any t not contained in the wavefunctional would be a non-dynamical time, disconnected from the dynamical system described by the wavefunctional. But the basic postulates of the theory tell us there can be no time parameter. Hence the right hand side of the equation is 0. Rather than expressing the dependence of the state on an external time, as in the Schrödinger equation, the constraint expresses only the fact that the quantum state has no dependence on such an external time.

A full specification of the theory is given by the choice of configuration space, Hamiltonian (and possibly other) constraints. There is no place for a time parameter; time has truly been eliminated in the theory. Nor does time necessarily show up in the form of any solutions: many solutions are known[15] that have nothing like a time parameter.

There may be approximate notions of time which arise from properties of the solutions or observables, but no notion of time is needed to construct the theory. If a theory formulated along these lines is correct, time has disappeared from the fundamental notions needed to describe the physical universe.

3. Challenges to the argument for the elimination of time

We now turn to several challenges which have been made recently to the argument just sketched. To my knowledge these are new and in my opinion they deserve careful consideration. If they are right then not only is time still a necessary concept, we will have to find a different way to frame dynamical laws. The proof of these challenges then cannot depend only on whether or not they point up a failure of the argument for the elimination of time. Their contribution must be at least equally positive as negative, they must point us towards the invention of a new dynamical framework in which time plays a different, and more essential, role than at present.

The arguments for the elimination of time as a concept necessary for the expression of the fundamental laws of physics follows from the five postulates mentioned above. If they all hold then the argument goes through. At a purely formal level, they do seem to hold for quantum general relativity in 3+1 and more dimensions, coupled to any kind of matter. The progress of the last 16 years using the Sen-Ashtekar formalism[16] and loop quantum gravity[17] strengthens the argument as it leads to the explicit construction of solutions to the Hamiltonian constraints.[18] Supersymmetry seems to make no difference, nor is there any reason to believe that they are not satisfied for string theory, even given that we do not know the framework of string theory at the background independent or non-perturbative level.

If the argument goes through then there seems little alternative to agreeing with the point of view expressed eloquently by Julian Barbour in *The End of Time*. This is the end of the concept of time in fundamental physics.

There seems only one real hope of evading the argument, which is to demonstrate that there is in fact no way to realize all five principles on which the argument depends consistently in a single theory. Is this possible? It is certainly the case that there are consistent realizations of *models of quantum theories of cosmology* in terms of models with finite numbers of degrees of freedom. These include quantum gravity in 1+1 and 2+1 dimensions as well as some models of very symmetric universes. However these only satisfy postulates **B, D** and **E** because they are too simple to contain subsystems complex enough to be called an observer. It is then an open question whether there are any theories which satisfy all of the postulates.

3.1 A first challenge: are there observables without time?

Postulate **C** requires the construction of a sufficient number of observables of the theory to distinguish the solutions from each other. As we are dealing with a theory with an infinite number of degrees of freedom this means we must have an infinite number of observables. In the introduction we discussed some of the issues involved with the construction of observables in general relativity. To sharpen this discussion we may distinguish two possible approaches to the construction of observables in classical and quantum theories of gravity:

- **Causal observables.** These are instructions for the identification of observables that make explicit reference to the causal structure of a classical or quantum spacetime. Since the causal structure is a diffeomorphism invariant of the metric, such an observable may be diffeomorphism invariant by construction. Examples are known which are of the following form: Identify a particular localized system as a local reference system and identify one of its degrees of freedom as a clock. Define observables in terms of the values of other of the degrees of freedom coincident with the clock variable taking on particular values. These correspond to actual observations that could be made in a spacetime, which, using the causal structure, give information about the spacetime to the past of the event when that local clock variable had a particular value.
  Such an observable can be constructed explicitly in a history's formulation of the theory without solving any additional conditions. They can in principle be directly implemented in a path integral formulation of the theory based on summing over Lorentzian histories.
- **Hamiltonian constraint observable.** These are observables which are constructed according to the rules of the hamiltonian formulation for systems with time reparameterization invariance. They must do at least one of the following things; (i) have vanishing Poisson bracket with the classical hamiltonian constraint; (ii) be expressed directly as a functional on the reduced phase space which is the constraint surface mod gauge transformations; (iii) commute with the quantum hamiltonian constraint.

Observables of the first kind make explicit use of the causal structure and hence use time in an essential way in their construction. If we only work with these observables then we have not eliminated time from the theory. Furthermore, there are no obstacles to defining and working with such observables, since no equations need to be solved, as they are diffeomorphism invariant by construction.

On the other hand, if we eliminate time from the theory, as sketched above, by either reducing the classical theory to its reduced phase or configuration space or constructing a quantum theory from solutions to the Hamiltonian constraint, then we have available only the second class of observables. We then conclude that Postulate **B** requires that we be able to construct an infinite number of observables of the second kind.

The Present Moment in Quantum Cosmology

The problem is that Hamiltonian constraint observables are extremely difficult to construct in real field theories of gravitation. There are formal proposals for how to construct such observables, which have been implemented with toy models. But these toy models are too simple and do not have local degrees of freedom that could be identified with fields measured by local observables inside the spacetime, or with such observables themselves. No observables have ever been constructed through any of the three methods mentioned which are local in the sense that they correspond to what an observer inside the universe would see.

The problem that arises in the first method is that to give such an observable explicitly as a functional on the phase space of the theory requires inverting the equations of motion of the theory over the whole space of possible initial data. Similarly, the second method has not been implemented because no one has a proposal for how to actually construct the reduced phase space. This also would involve an inversion of the equations of motion of the theory. The problem with the third approach is that while we have been fortunate enough to find an infinite number of quantum states which are exact solutions to the Hamiltonian constraints of theories of gravity,[19] finding operators which commute with the constraints has proved to be much harder, and only a few, rather trivial, examples of such operators have been found. No one has even proposed a practically implementable strategy for how to construct operators that both commute with the quantum Hamiltonian constraint and refer to local observations.

So whether or not there exist in principle observables of the second kind, there are no known methods to construct them for realistic theories.

3.2 Newman's worry: the implications of chaos

Ted Newman[20] and others have raised a further argument against the existence of observables of the second kind, constructed through either method (i) or (ii) which is that the fact that gravitational systems have chaotic behaviour may in principle prevent their construction. It is well known that the motion of three bodies under their mutual gravitational attractions is chaotic in Newtonian mechanics. It is then hard to believe that the typical solution of general relativity is not also chaotic. It is also known that the generic cosmological model with a finite number of degrees of freedom, called the Bianchi IX model, is chaotic, at least if a physically realistic notion of time is used in the criteria for chaos. If we do not yet have a proof that the Einstein equations are chaotic it is mainly because of the problem of finding a definition of chaos suitable for the application to a field theory of gravity.

But if the equations are chaotic it means that physical observables are not going to be representable by smooth functionals on the phase space. The reason is that they must be constant under the trajectories defined by the hamiltonian constraints, as these must be satisfied by any physical, diffeomorphism invariant observable. But if the system is chaotic then any trajectory passes through any open set in the phase space. This means that physical observables will take an infinite number of values in any open set. It is easy to construct examples of such observables which are not continuous anywhere inside at least certain open sets.

It will be difficult both to construct such observables explicitly and to evaluate them. It may also not be possible to define an algebra of such observables, as this requires taking derivatives to form the Poisson bracket. But if such observables do not have an algebra, they cannot be represented in the quantum theory, as the quantum theory is generally defined in terms of a relationship between the algebra of quantum operators and the Poisson algebra of classical observables.

Newman's worry turns what may seem at first only a practical problem into a problem of principle. It may be not only beyond our power to construct observables for a formulation of cosmology in which time has been eliminated, it may be impossible in principle. If this is the case then the argument fails, for a theory without observables that can be made to correspond to what we real observers can measure is not a physical theory, it is just a formal structure with no possible relation to the real world.

3.3 Markopoulou's argument:
the configuration of the universe is not observable

In a recent article Markopoulou[21] pointed out that postulate **C**, observability of the configuration space, is not likely to be satisfied by any quantum theory of gravity whose classical limit is general relativity. The reason is that in generic cosmological solutions to classical general relativity containing dust and radiation, with spacetime topology of the form $\Sigma \times R$, with Σ either R^3 or S^3, the backwards light cones of typical events do not contain the full Cauchy surface Σ. This means that Postulate **C** fails in these examples.

It is possible to avoid this conclusion in several ways. The first is to have a 'small universe' in which the spatial topology is more complicated. Presently this appears to be ruled out by observation. The second is to weaken the requirement that the matter be described by dust and radiation. By doing so we can allow inflation to have occurred in our past. There are, however, two problems with saving the observability of the universe by inflation. The first is that without fine tuning inflation predicts that the universe is spatially flat. In this case we certainly do not see a whole Cauchy surface in our past. The second is that according to inflationary models the whole Cauchy surface would be seen only during the inflationary era when the quantum fields are close to their ground state. The inhomogeneities that drive structure formation in our universe are hypothesized to arise only at the end of the inflationary period, in a 'quantum to classical transition' in which quantum fluctuations in the vacuum state are converted to classical fluctuations in the matter and geometry. This transition is believed to be akin to a measurement of the quantum state of the fields. The result is that the information necessary to determine the classical geometry of the universe is nevertheless not available in the backwards light cone of any one observer.

More could be said about these discussions, but the conclusion is that Postulate **C** is likely not satisfied in our universe.

We are then faced with the following choice: Give up postulate **C** and accept a classical or quantum theory of cosmology which is formulated in terms of quantities that are not observable or attempt to modify the theory so as to formulate it only in terms of quantities observable by real observers inside the universe. Markopoulou[22] describes a framework in which general relativity and other spacetime theories may be reformulated completely so that the only observables involve information available to internal observers by means of information reaching them from their backwards light cones.

The potential power of Markopoulou's argument rests on the following simple observation: in classical general relativity the causal structure plays a significant role in delineating what is observable. To begin with Markopoulou points out that if we insist on the principle that any observable of a cosmological theory must in fact be observable by an observer inside the universe then it follows that our theory must be pluralistic in the particular sense that different observers have access to different information. However this is a structured pluralism because the causal structure implies certain logical relations amongst the observations made by different observers. Markopoulou then points out that the pluralistic logic required to define what is observable in general relativity is closely related to logics studied by mathematicians under the name of topos theory. The basic rules that relate observations made by different observers follow from the causal structure and the requirement that whenever two observers receive information from the same event they will agree on the truth-value given to any propositions about that event.

Markopoulou's argument then formalizes the worry about novelty I mentioned in the introduction. The key point is that since any given observer, located in some finite region of space and time, only receives information from a proper subset of the events in the history of the universe; and furthermore, since no observer sees a whole Cauchy surface; they have no way to predict what information will be received from systems of the universe they will come into causal contact with at a later time. As a result, the logic of observables in a cosmological theory must not satisfy the law of the excluded middle. No single observer can assign the values **TRUE** or **FALSE** to all of the propositions which describe the history of the universe. There must be other truth-values such as, **cannot yet be determined by this observer**. It turns out that one has to be able to describe different degrees of ignorance, so that there is actually a whole algebra of possible truth-values besides true or false.

The result is that the logic of observables in a relativistic cosmological theory must be both non-boolean and pluralistic. For any single proposition, different observers will be able to assign different truth-values. Whenever two are able to assign TRUE or FALSE they must agree. But they need not agree if one or both are only able to assign truth-values which indicate some amount of inability to determine.

So far we have been discussing only the classical theory. There are also implications for the quantum theory which Markopoulou develops in further papers.[23] The point is that the possible truth-values of the classical theory must appear in the spectrum of projection operators in the quantum theory. But, she points out, one cannot realize such a non-Boolean pluralistic logic in terms of the spectrum of operators on a single Hilbert space. Consequently, a quantum theory of cosmology whose classical limit reproduces the algebra of observables of the classical theory explored by Markopoulou cannot be based on a single Hilbert space. As Markopoulou describes,[24] it can, however, be constructed from a presheaf of Hilbert spaces, which is a structure in which there is a Hilbert space assigned to every event in a causal set.

The result is a new framework for quantum cosmology in which the causal structure, and hence time, plays a fundamental role in the formulation of theory. The basic implication of Markopoulou's work for the present argument is then that to the extent that the causal structure plays a role in the identification of observable quantities in general relativity, it must also play an essential role in the construction of the corresponding

The Present Moment in Quantum Cosmology

quantum theory. If we are then restricted to the first kind of observables, which are causal observables, then time cannot be eliminated from quantum cosmology.

3.4 Kauffman's argument: the configuration space cannot be described in advance of the evolution of the system

We come now to a challenge to the argument that attacks the first postulate, which is the constructibility of the configuration space.[25] This remark was inspired by the work of Stuart Kauffman on theoretical biology and economics in which he raises two very interesting questions:[26]

- *Is it possible that the configuration relevant for mathematical biology, ecology or economics cannot be constructed by any finite mathematical procedure?*
- Even if the answer to the first is in principle no, is it possible that the construction of the relevant configuration spaces are so computationally intensive that they could not be carried out by any subsystem of the system in question?

There are good reasons to suspect that the answer to at least one of these questions is yes when we are dealing with such potentially large and complex configuration spaces as the space of all possible phenotypes (as opposed to genotypes) for biological species, the space of all properties that might be acted on by natural selection, the space of all biological niches, or the space of possible kinds of businesses or ways of earning a living.

We refer the reader to Kauffman[27] for discussions of the implications of this possibility for theories of biology and economics. For the present purposes we are interested in considering the analogous questions:[28]

- Is it possible that there is no finite procedure by means of which the configuration space of the universe may be constructed?
- Even if the answer is no is it possible that the computation that would be required to carry out the construction of the configuration space is so large that it could not be completed by any physical computer that existed inside the universe?

The possibility that the answer to one or both questions is yes arises from the fact that, as pointed out in section 2 above, the physical configuration spaces relevant for cosmological theories are quotients of infinite dimensional spaces by the action of infinite dimensional groups. For the case of general relativity in 3+1 or more dimensions the physical configuration space is defined to be the quotient of the space of metrics on some fixed compact manifold, Σ by the diffeomorphism group of Σ. No closed form representation of the quotient is known, even in the simplest case in which $\Sigma = S^D$. The space of metrics on a manifold Σ is known to be an infinite dimensional Riemannian manifold. Coordinates are known which cover it and the tangent space, metric, connection and curvature tensor are known. So the issue is not just the infinite dimensionality. The problem is that the diffeomorphism group is very large and its action is quite complicated. It is known that the quotient has singularities and is not everywhere an infinite dimensional manifold.

To define the quotient we must also be able to have a procedure to answer the following question: given two metrics, g_{ab} and g'_{ab} on Σ, does there exist a diffeomorphism $\phi \in Diff(\Sigma)$ such that $g'_{ab} = \phi \circ g_{ab}$? By computing a large enough set of curvature scalars one can reduce this problem to the question of when there is a diffeomorphism that can bring one set of functions on a manifold to another. But no finite procedure is known which can answer this question in general even in the case that we restrict to the analytic category in which both metrics and diffeomorphisms can be defined by power series expansions in some set of coordinates. The problem is that while one may be able to show in a finite number of steps that g_{ab} and g'_{ab} are not diffeomorphic, there is no finite procedure which works in the general case to tell whether they are or not.

But if one cannot tell whether two metrics are in the same diffeomorphism class or not one is not going to be able to construct the quotient by an explicit construction of the equivalence class within the space of metrics on Σ. If this is the case, postulate **D** is not satisfied.

One might try to avoid this by use of the alternative Postulate **D'**, which replaces the requirement of constructing the space of metrics mod diffeomorphisms by the solution of certain functional differential equations. This is suggested by the fact that in one approach to quantum gravity a complete solution to the problem of solving the diffeomorphism constraints has been stated in closed form. This is the loop quantum gravity approach.[29] The basic result of loop quantum gravity is a precise recipe for the construction of an orthonormal basis for the Hilbert space of quantum general relativity on a spatial manifold with topology and differential structure Σ. The basis is in one to one correspondence with the diffeomorphism equivalence classes of the embeddings in Σ of a certain class of labelled graphs called spin networks. The problem we are

The Present Moment in Quantum Cosmology 133

concerned with then reduces to the question of classifying the diffeomorphism classes of embeddings of these graphs.

There is no problem enumerating and distinguishing finite labelled graphs as this comes down to distinguishing finite dimensional matrices with integer entries. However, classifying and distinguishing diffeomorphism classes of the embeddings of graphs in three dimensions is a tricky question. The problem of the classification of ordinary knots up to ambient isotopy has only recently been solved and even that requires a very computationally intensive procedure which involves classifying the kinds of groups that appear as the homotopy group of the complement of the knot. The homotopy groups of complements of graphs present a greater challenge, and, to my knowledge, these remain unclassified.

Even if this problem is solved the problem of distinguishing diffeomorphism classes is a good deal trickier for graphs with arbitrary valence intersections than the problem of distinguishing ambient isotopy classes. The reason is that in the case of sufficiently high valence (5 and higher in three spatial dimensions) the diffeomorphism invariance classes are labelled by continuous parameters.

If it turns out that there is no finite procedure for solving either problem then it follows there will be no finite procedure to construct the Hilbert space for quantum general relativity and related theories (including all known couplings to matter, including supersymmetric theories).

The problem is made even more complicated if one believes, as many do, that the topology of Σ should also be able to change by means of quantum transitions. In this case one has to classify networks embedded in arbitrarily complex three topologies. But it is sufficient to note that the problem is already quite serious for a fixed simple topology.

Finally, we note that even if the answer to the first question is no, the answer to the second is certainly yes. The maximum computational power of a universe could not increase faster than its volume, which is roughly proportional to the number of nodes of the graph of the corresponding quantum state. But the number of steps required to distinguish the embeddings of two graphs is likely to go up faster than any power of the number of nodes. We may also note that the situation is likely worse than this if the holographic principle is correct, as that bounds the amount of information a region of space could contain to its area rather than its volume in Planck units.

To summarize this part of the argument: it is quite possible that the answer to the first question raised above is yes, and almost certain that the second is. In the case that the configuration space cannot be constructed, the argument for the elimination of time in classical cosmology cannot be run, for the whole framework falls apart without a prespecified configuration space. Similarly, if the Hilbert space of states cannot be constructed, the argument for the elimination of time in quantum cosmology cannot be run.

In either case, if the spaces are constructible, but require more computational power than the universe could contain, then we are faced with an interesting situation. A mythical extra-universe observer could run the argument for the elimination of time. But this is impossible for any real observers inside the universe. A quantum theory of cosmology that requires more processing power than the universe could contain to set up its Hilbert space and check whether two states were orthogonal or not would not be a theory that we who live inside the universe could use to do real computations. So it is not clear what relevance for our physics there may be for the possibility that some imagined being outside the universe could eliminate time from physics. What matters is that we cannot work with a physical theory without time.

4. Conclusions

Let me begin my summary of these arguments by emphasizing the role played by the requirement that a theory of cosmology must be falsifiable in the usual way that ordinary classical and quantum theories are. This leads to the requirement that a sufficient number of observables can be determined by information that reaches a real observer inside the universe to determine either the classical history or quantum state of the universe. Only if this is the case can we do cosmology within the standard ideas concerning the methodology and epistemology of dynamical theories.

It does appear that this is not the case. Interestingly enough, this statement requires input from both theory and observation. General relativity allows both possibilities. Presently observations seem to rule out the possibility that we live in a universe in which there is a complete Cauchy surface within the classical region of the past light cones of typically situated observers such as ourselves.

The implication seems clear: if we want to do cosmology as a science, we must restrict ourselves to theories in which all observables are accessible to real observers inside the universe. To do this we must invent a new framework for quantum cosmology which does not include notions like 'the wavefunctional of the universe'. One way to do this has been proposed by Markopoulou, which is called 'quantum causal histories'.[30] These are an interconnected web of Hilbert spaces tied to the causal structure in such a way that each act of observation, considered as a particular event in the history of the universe, is represented in terms of a Hilbert space constructed to represent the information available to be observed at that event. These together provide a representation of projection operators whose spectrum consists of the observer dependent truth-values we discussed above.

This proposal may be seen in the light of a general issue which has been discussed a great deal in quantum cosmology, which is that of context dependence. This arises most generally in the consistent or decoherent history approaches to quantum cosmology.[31] As pointed out by Gell-Mann and Hartle,[32] and Dowker and Kent,[33] the consistent histories formulation requires the specification of a context within which observations are to be made, prior to their interpretation. This is necessary to replace the quantum world/classical world division of Bohr's interpretation of quantum theory in a way that avoids the preferred basis problem of the Everett interpretation. Isham and collaborators[34] have studied the general mathematical structure of such contextual approaches to cosmology and found they are naturally formulated in terms of topos theory. Markopoulou's quantum causal histories can be understood from this point of view as the result of using the causal structure of the history of the universe to define the contexts.

However, the general issue of context dependence does not by itself refute the argument for the elimination of time. Were it possible for a single observer inside the universe to make sufficient measurements to determine either its classical history or its quantum state, and assuming that the other four postulates also held, the argument for the elimination of time could be run. In this circumstance it might still be convenient to express quantum cosmology in terms of histories, but it would not be essential. Barbour and others would be able to argue that they could do cosmology perfectly well with no notion of histories apart from what was necessary to recover the classical limit.

It is only by insisting that the context of real observers inside their universe is defined by the information that reaches them by means of radiation that propagated from their past that a link is made between the issue of observability of the universe and its causal structure. Of course, the notion of causal structure may be loosened quite a bit from that which arises in general relativity, as has been done in various causal set and evolving spin network models of quantum gravity. But it is hard to divorce the notion of causal structure from the idea that there is a finite speed for the propagation of information, and hence from some notion of time.

Is the notion of time then built into the argument from the beginning? No, the key point is the insistence on building a cosmological theory that makes references only to observations made by real observers inside the universe. It is then an observed fact that the universe is very big compared to its observers. A combination of observation and theory then leads to the conclusion that the observations made by one observer at one 'moment' are insufficient to determine the classical or quantum state of the whole universe. We may note that all that is required here is the notion of a moment at which a number of simultaneous measurements may be made. This is already all we need to argue that cosmology requires a different framework from a conventional quantum system because the postulates of quantum theory require that the state of a quantum system must be uniquely determined by measurements of a complete set of commuting observables which, by definition, can be made simultaneously.

This is already sufficient to refute the argument for the elimination of time. The proposal of Markopoulou builds up from here to propose an alternative way to do cosmology, which is in terms of a structured set of observations that are not made at simultaneous moments. The positive proposal is that the structure which is imposed on the possible measurements is a partial ordering which is derived from the causal structure of the universe.

This is a good moment to recall Newman's worry. This may be put in the following form. Even if the universe is governed by classical deterministic equations, these are likely to be chaotic. This is important because of the fact that all measurements we real observers can make are of finite precision. It means that, even if all five postulates are true, and the argument for the elimination of time can be run, we still will be able to make no reliable predictions of measurements that refer to 'moments' when our clocks read different 'times'. Even if time can be eliminated in principle, it cannot be eliminated in practice.

There is a stronger form of Newman's worry, which has to do with the existence of a physical observable algebra for a chaotic but time reparametrization invariant system. But even leaving this aside we see that this argument has an interesting parallel with Markopoulou's argument, since both refer to the information available to a real observer inside the universe. They concern two ways in which real observers differ from

The Present Moment in Quantum Cosmology

abstract idealized observers in the real universe. First, that they are small and are limited to information available in a local region of spacetime. Second, they can only make observations of finite precision. It is very interesting that both arguments suggest it may be useful to think about measurement in information theoretic terms. In particular both suggest it is important to make a distinction between what can be done in principle in a mathematical model and what can be done in a finite number of steps.

A related issue of constructibility underlies the third argument, introduced by Kauffman. The implication of this argument is that if we want to have a measurement theory which is relevant to what real observers inside the universe do we must require that all the constructions necessary to formulate and compute in the theory can be carried out in a number of operations which is not only finite but small enough to be carried out by an observer inside the universe.

In the cases of theoretical biology and economics, Kauffman goes on to propose a method for formulating a theory that does not depend on the constructibility of huge and complex configuration spaces. He proposes that it may be sufficient that the theory provide an algorithm that allows the construction of all configurations which differ by a small number of local changes from any given present configuration. This space of possible nearby configurations is large enough for a large complex system, but still infinitely smaller than the space of all possible configurations. Kauffman calls it *the adjacent possible*.

Could we do quantum cosmology in terms of such an adjacent possible? A start on such an approach is made in 'Combinatorial dynamics in quantum gravity'. Suppose that the state space of a quantum theory of gravity were not constructible. For example suppose this is in fact the case for the ambient isotopy classes of embeddings of spin networks in a three manifold. It is still possible to construct the space of all spin networks that differ from any given one, Γ, by a local one of a small set of possible local moves. The local moves can be moves that involve rearrangements of less than P nodes, which together with their common edges must make an n simplex for $n \leq P$ These are the forms of moves appropriate to P spacetime dimensions. Given Γ let Ω_Γ^r be the linear space spanned by all spin networks that are the result of r local moves made on Γ. This will be a finite dimensional space. We may now use the fact that the evolution moves in theories of this type are themselves local moves of the same type. If we use the idea that each local change constitutes the analogue of an event in a quantum spacetime[35] together with the idea that each event will correspond in the classical limit to a small, but finite, spacetime volume, we can see that the whole space of quantum universes which begin with the initial state Γ and have finite spacetime volume, V, in Planck units, will live in some Ω_Γ^r, with r a finite function of V. But for any finite r and finite Γ, Ω_Γ^r is a finite dimensional constructible space. It then appears that the answer to the question is yes, quantum cosmology could be formulated in terms of an adjacent possible type construction.

However, the resulting theories are still subject to the argument of Markopoulou: the states in Ω_Γ^r are not measurable by any observer inside the universe. To take both into account we should realize this kind of construction in a quantum causal history. This is in any case a natural thing to do as a quantum causal history is based on a partial order structure, and such a structure is naturally generated by local moves made on graphs. The result is a framework for quantum cosmology which escapes the worries raised here.

In these theories the notion of information plays a crucial role. This is forced on us by the combination of finiteness of the space of local changes and finiteness of the propagation of the effects of local changes. But it also arises from an apparently independent set of considerations having to do with the Bekenstein bound[36] and the holographic principle.[37] These suggest that measures of geometry may actually be reduced to measures of information flow. In its 'weak form', the holographic principle asserts that the geometric area of any surface must be reducible, in a fundamental theory, to a measure of the capacity of that surface as a channel of flow of information from its causal past to its causal future.[38]

In closing we note the very interesting way in which notions of finiteness and constructibility are coming into fundamental theories of quantum cosmology. This may not be surprising to experts in philosophy and the foundations of mathematics, for whom these notions are closely related to ideas on time. But it is a new idea for some of us who come at it from physics: if the universe is discrete and time is real, and is itself composed of discrete steps, then time may be none other than the process which constructs, not only the universe, but the space of possible universes relevant for observations made by local observers. Beyond this, there is the possibility of a quantum cosmology in which the actual history of the universe up till some moment and the space of possible universes present at that 'instant' are not two different things, but are just different ways of seeing the same structure, whose construction is the real story of the world.

Centre for Gravitational Physics and
Geometry,
Department of Physics,
The Pennsylvania State University,
University Park, PA, USA 16802

The Blackett Laboratory,
Imperial College of Science,
Technology and Medicine
South Kensington,
London SW7 2BZ, UK

Notes

1 Acknowledgements: This paper is mainly a commentary on ideas of Kauffman, Markopoulou and Newman, as indicated in the introduction. I would like to thank the theoretical physics group at Imperial College for their hospitality during this last year. This work is supported by the NSF through grant PHY95–14240 and a gift from the Jesse Phillips Foundation. The title of the paper is taken from a debate with Jaron Lanier, held in Brooklyn, New York, April 3, 1999, under the auspices of Universita Universalis.
2 Julian Barbour, *The End of Time* (London: Weidenfeld & Nicolson, 1999).
3 Julian Barbour, *The Timelessness of Quantum Gravity: I. The evidence from the classical theory; II. The appearance of dynamics in static configurations*, in *Classical and Quantum Gravity*, 11 (1994), 2853.
4 Stuart Kauffman, *Investigations* (Oxford: Oxford University Press, in press); Stuart Kauffman & Lee Smolin, 'Combinatorial Dynamics in quantum gravity', hep-th/9809161 (*Proceedings of the Karpaz 99 winter school,* 1999).
5 F. Markopoulou, 'The Internal logic of causal sets: What the universe looks like from the inside' (1999); gr-qc/98011053,* to appear in *Commun. Math. Phys.* (2000); Markopoulou, 'Quantum causal histories' (1999), hep-th/9904009,* to appear in *Class. Quantum Grav.* (2000); Markopoulou, 'An insider's guide to quantum causal histories', hep-th/9912137.*

  * Citations in these formats indicate references to the archive of papers kept at http://xxx.lanl.gove – 'a fully automated electronic archive and distribution server for research papers. Covered areas include physics and related disciplines, mathematics, non-linear sciences, computational linguistics, and neuroscience.'
6 E.T. Newman , personal communication.
7 A diffeomorphism is in this context a map of a manifold to itself that preserves the notion of infinitely differentiable functions. Thus, it moves the points around, but in a way that takes differentiable functions to differentiable functions. It thus preserves relationships between functions that can be described by coincidences of values at points.
8 For example in asymptotically flat or DeSitter spacetime one can define time evolution with respect to a time coordinate at infinity. There is a problem with how to continue this into the interior of the spacetime, but this is not the same as the problem which occurs in cosmological theory. It can be

9   Markopoulou, 'The Internal logic of causal sets: What the universe looks like from the inside'; 'Quantum causal histories'; 'Dual formulation of spin network evolution' (1997), gr-qc 9704013.
10   See, for instance, *The Leibniz-Clarke Correspndence*, trans. & ed. R. Ariew (Indianapolis: Hackett, 2000); Leibniz, Letter 3, §4, and *passim*.
11   Motion on such relative configuration spaces was first studied by Barbour, and Bertotti and Barbour.
12   This is a key step of the argument; for the complete version I refer the reader to the works by Barbour cited in endnotes 2 and 3, and Carlo Rovelli, 'Time in Quantum Gravity: Physics beyond the Schrödinger regime', *Phys. Rev.* D43.442 (1991).
13   Details can be found in Barbour, *The End of Time*, and Rovelli, 'Time in Quantum Gravity'.
14   For reviews, see: L. Smolin in *Quantum Gravity and Cosmology*, eds. J. Pérez-Mercader *et. al.* (Singapore: World Scientific, 1992); 'The Future of Spin Networks', gr-qc/9702030; R. Gambini & J. Pullin, *Loops, Knots, Gauge Theories and Quantum Gravity* (Cambridge: Cambridge University Press, 1996); C. Rovelli, 'Loop Quantum Gravity' (1997), gr-qc/9710008.
15   Rovelli, 'Loop Quantum Gravity', Gambini & Pullin, *Loops, Knots, Gauge Theories and Quantum Gravity*; T. Jacobson & L. Smolin, 'Non-perturbative quantum geometries', in *Nuclear Physics* B299 (1988), 295–345; C. Rovelli & L. Smolin, *Phys. Rev. Let.* 61 (1988), 1155; Rovelli & Smolin, *Nucl. Phys.* B133, 80 (1990); Rovelli & Smolin, 'The physical Hamiltonian in non-perturbative quantum gravity', *Phys. Rev. Lett.* 72 (1994), 446; Rovelli & Smolin, 'Discreteness of area and volume in quantum gravity', *Nuclear Physics* B442 (1995), 593. Erratum: *Nucl. Phys.* B456 (1995), 734.
16   A. Sen, 'Gravity as a spin system', *Phys. Lett.* B119 (1982), 89; Sen, 'Quantum theory of a spin 3/2 system in Einstein spaces', *Int. J. of Theor. Phys.* 21 (1982), 1; A. Ashtekar, *Phys. Rev. Lett.* 57 (1986) 2244; Ashtekar, *Phys. Rev.* D36 (1987), 1587.
17   See papers cited in note 14.
18   See papers cited in note 15.
19   See papers cited in note 15.
20   E.T. Newman, personal communication.
21   Markopoulou, 'The Internal logic of causal sets: What the universe looks like from the inside'.
22   F. Markopoulou, 'Dual formulation of spin network evolution' (1997), gr-qc/9704013.
23   Markopoulou, 'Quantum causal histories'; 'An insider's guide to quantum causal histories'.
24   Markopoulou, 'The Internal logic of causal sets: What the universe looks

like from the inside'; 'Quantum causal histories'
25 This kind of worry goes back at least to a paper by Hartle in which he points out that the impossibility of classifying four manifolds is a problem for the Euclidean path integral formulation of quantum cosmology. See J. Hartle, 'Scientific Knowledge from the Perspective of Quantum Cosmology' (1996), gr-qc/9601046.
26 Kauffman, *Investigations*.
27 Kauffman, *Investigations*.
28 Kauffman & Smolin, 'Combinatorial Dynamics in quantum gravity'.
29 Loop quantum gravity is a mathematically well defined, non-perturbative and background independent quantization of general relativity, with its conventional matter couplings. The research in loop quantum gravity forms today a vast area, ranging from mathematical foundations to physical applications. Among the most significant results obtained is an intriguing physical picture of the microstructure of quantum physical space, characterized by a polymer-like Planck scale discreteness. See Rovelli, 'Loop Quantum Gravity', Gambini & Pullin, *Loops, Knots, Gauge Theories and Quantum Gravity*.
30 Markopoulou, 'Quantum causal histories'; 'An insider's guide to quantum causal histories'.
31 Griffiths, 'Consistent Histories and the Interpretation of Quantum Mechanics', *J. Statist. Phys.* 36 (1984), 219; Omnes, 'Logical Reformulation of Quantum Mechanics', *J. Statist. Phys.* 53 (1988), 893.
32 M. Gell-Mann & J. Hartle, 'Equivalent Sets of Histories and Multiple Quasi-classical Realms' (1994), gr-qc/9404013; J.B. Hartle, 'Generalized Quantum Theory and Black Hole Evaporation' (1998), gr-qc/9808070.
33 F. Dowker & A. Kent, 'On the Consistent Histories Approach to Quantum Mechanics', *J. Statist. Phys* 82 (1996), 1575; gr-qc/9412067.
34 J. Butterfield & C. Isham, 'Some Possible Roles for Topos Theory in Quantum Theory and Quantum Gravity' (1999), gr-qc/9910005.
35 Markopoulou, 'Dual formulation of spin network evolution'
36 J.D. Bekenstein, *Lett. Nuovo Cim.* 4 (1972), 737–740; *Phys. Rev.* D7 (1973), 2333–2346; *Phys. Rev.* D9 (1974), 3292–3300.
37 G. Hooft, 'Dimensional reduction in quantum gravity' (1993), gr-qc/9310006; L. Susskind, 'The world as a hologram', *J. Math. Phys.* 36 (1995), 6377–6396, hep-th/9409089; L. Crane, 'Categorical Physics' (1993), hep-th/9301061; Crane, 'Topological Field Theory as the Key to Quantum Gravity' in J. Baez (ed.), *Knot Theory and Quantum Gravity* (Oxford: Oxford University Press, 1993), hep-th/9308126; Crane, 'Clock and Category: Is Quantum Gravity Algebraic?', *J. Math. Phys.* 36 (1995), 6180–6193, gr-qc/9504038; F. Markopoulou & L. Smolin, 'Holography in a quantum space-time' (1999), hep-th/9910146; L. Smolin, 'The strong and weak holographic principles' (2000), hep-th/003056.

38 Markopoulou & Smolin, 'Holography in a quantum space-time'; Smolin, 'The strong and weak holographic principles'.

Glossary

Diffeomorphism invariance The mathematical expression of the principle that space and time are an expression simply of the relationships between events in the world.

Equivalence class (of manifolds or metrics) is the class of all members of a manifold or metric that are in a given equivalence relation, where equivalence relation denotes a reflexive, symmetric or transitive relation between elements of the manifold or metric.

Field The means by which a force communicates its influence, described by a collection of numbers at each point in the field that reflect the strength and direction of the force at that point.

Gauge principle The basic principle behind the 'standard model' of elementary particle physics, which expresses the idea that all physical properties of elementary particles are defined in terms of interactions between them. The standard model unifies electromagnetic, weak and strong interactions, but not gravity.

Gauge invariance The property of electrically charged particles in quantum theory stating that the phase-angle (one of the two elements of the wave-function, along with the amplitude) has no physical significance, and can be chosen at will at each point of space.

Gauge transformation The introduction of a gauge function into an equation, often for symmetry purposes. A gauge function that is introduced into an equation does not alter the derived equations of observable physical quantities.

Hamiltonian function A Hamiltonian function is a function used to describe a system of particles in terms of their positions and momenta.

Hilbert space A normed space having an infinite number of dimensions with each point at a finite distance from the others.

Homotopy The property characterising a mapping that deforms one path continuously into another, such that all intermediate paths lie within the topological space of which the two given paths are subspaces.

Inhomogeneity A thing that is not homogeneous with its surroundings, which is locally irregular; in mathematics, an inhomogeneity consists of terms that are not all of the same degree or dimensions.

Isotopy Describing a class of knots: two knots are isotopic if it is possible to get one from the other by applying a diffeomorphism of space that can be continuously deformed to the identity.

Manifold A continuous space such as a real number line; a **metric** of a manifold is a mathematical description of the geometry of the manifold.

Path integral formulation One of two approaches to quantum mechanics – the other being the canonical, or Hamiltonian formalism – path integral formalism was introduced by Richard Feynman, and developed by James Hartle and Murray Gell-Mann.

Phase One of several possible different configurations of some material; more generally, phase refers to the possible descriptions of a physical system as the features on which it depends are varied; a **phase space** is a space whose coordinates are given by the set of independent variables characterising the state of a dynamical system.

Quantum gravity Theory which seeks to unify in one framework both quantum and gravitational physics.

Spin Intrinsic property of certain elementary particles which is a form of angular momentum, usually represented as a rotation.

Spin network A simple model of discrete quantum geometry, first introduced by Roger Penrose, it is a system consisting of a number of 'units', each of which has a total angular momentum, and which interact in ways that conserve angular momentum. The system is purely relational, since it makes no reference to background notions of space, time or geometry. Without a background geometry, a particle can only have a total angular momentum, as there is nothing with respect to which a direction in space may be defined.

Supergravity A class of point-particle theories combining general relativity and supersymmetry.

Supersymmetry A symmetry principle that relates the properties of particles with a whole number amount of spin (bosons) to those with half a whole (odd) number amount of spin (fermions).

Symmetry A property of a physical system that does not change when the system is transformed in some manner.

Wave function The mathematical object in quantum mechanics that determines the probability density of a particle in space and time, a complex quantity comprising the amplitude (whose square gives the probability) and a phase-angle.

7
Duration and Evolution: Bergson contra Dennett and Bachelard

Keith Ansell Pearson

> Modern science dates from the day when mobility was set up as an independent reality. It dates from the day when Galileo, rolling a ball down an inclined plane, made the firm resolution to study this movement from high to low for itself, in itself, instead of seeking its principle in the concepts of the high and the low, two immobilities by which Aristotle thought he sufficiently explained its mobility. And this is not an isolated fact in the history of science. I take the view that several of the great discoveries, of those at least which have transformed the positive sciences or created new ones, have been so many soundings made in pure duration. The more living was the reality touched, the more profound had been the sounding.
> Bergson, *Introduction to Metaphysics*[1]

> The duration of the universe must therefore be one with the latitude of creation which can find place in it.
> Bergson, *Creative Evolution*[2]

How should the 'new' be thought? This question remains at the forefront of philosophical disputation and, interestingly, has once again taken the form of an encounter between Bergsonism and its critics, with the agon between Badiou and Deleuze echoing the complaints made in the 1930s by Bachelard contra Bergson. In contesting Deleuze's Bergsonism, for example, Badiou has argued in favour of the 'founding break' over 'creative continuity', in favour of the 'stellar separation of the event' over the 'flux'. If, in the occurrence of an 'event', there is a creative excess, then for Badiou, this creativity comes not from the inexhaustible fullness of the world but rather from the event not being attached to it *and* 'in the absence of continuity' (it comes from its being separated and interrupted).[3] And, just like the foremost critic of Bergsonism of a previous generation, Bachelard, Badiou has recourse to set theory and its selection of actual multiples as a way of contesting the reliance on virtual multiplicities. The dispute is not over the reality of the new, but precisely how to think the production or creation of the new. For Bachelard and Badiou the 'new' is, almost by definition, that which exceeds prior conditions and which cannot be explained in terms of them. The quarrel with Bergsonism appears to rest on the claim that the new cannot be genuinely new if it is bound up, in however complicated a fashion, with the past. Bachelard, for example, sought to reject completely Bergson's attachment to continuity because, it appeared to him, this meant that the present was always inscribed in the past: the 'solidarity of past and future' and the 'viscosity of duration' mean, he argues, that 'the present instant is never anything other than the phenomenon of the past'.[4] For Badiou the 'event' has no relation to time or duration, it is a punctuation in the void of being and a complete interruption in the order of time (if it can be given a temporality it is only of a retroactive kind).

For Bergson, and Deleuze following him, however, the new is bound up with a creative unrolling and enrolling, and cannot be conceived outside of duration. Contra Badiou, Deleuze argues that to think the new, or the event, otherwise is to reintroduce *transcendence* into philosophy. To talk of the production of the new in terms of an interruption or founding break is to render it mysterious and almost inexplicable. In this essay I want to demonstrate, contra the criticisms made of Bergsonism from Bachelard to Badiou, that duration is the condition of novelty. Duration is not the transcendent(al) condition of the new but the *immanent* condition of its *real* production. My focus will be on *Creative Evolution*, the text in which Bergson does not restrict duration merely to a phenomenological provenance but extends it to the entire cosmos. It is, I believe, through his response to the question of 'evolution' that the challenge of Bergson's thinking of duration can best be demonstrated. The challenge consists, in essence, in not reducing the present to the status of being little more than a mere brute repetition of the past – if it were, it would be difficult to see how time could be given in Bergson's conception of a creative evolution (there would be nothing 'creative' about it and there would not even be a phenomenon we could describe as 'evolution') – but rather in thinking a duration 'in which each form flows out of previous forms, while adding to them something new, and is explained by them as much as it explains them...' [CE 362/Œ 801].

What is duration? We follow Deleuze in maintaining that duration is not reducible to psychological experience but can be seen to be providing the theme of a complex ontology.[5] In addition, duration can also be approached in terms of a 'transcendental empiricism'. Although Bergson insisted upon the empirical character of both the *élan* vital and duration,[6] these notions also possess a transcendental status – 'transcendental' not in the sense of subjective conditions of possibility but rather in the sense of making a phenomenon intelligible, in this case 'evolution'. For Deleuze, one of the most important innovations of Bergson's philosophy is to be found in the way he thinks duration as a multiplicity. As he notes, multiplicity is not part of the traditional vocabulary of philosophy when denoting a continuum.[7] It is in *Time and Free Will* that Bergson first introduces the distinction between a discrete multiplicity and a continuous one, a distinction that Deleuze reworks in terms of actual and virtual multiplicities.[8] The difference is between thinking objects and things discretely, whereby the relations between them are ones of juxtaposition and exteriority, and thinking the components of a system in terms of fusion and interpenetration. This notion of a virtual multiplicity becomes important for thinking evolution since it serves to show that we do not have to oppose heterogeneity and continuity. Why is this so? Duration is not a simple indivisible which admits of no division; the point is rather that it changes in kind in the very process of getting divided up. This is why Deleuze treats it as a non-numerical multiplicity 'where we can speak of "indivisibles" at each stage of the division' (as in the run of Achilles).[9] But what is it, in the case of evolution, that is getting divided up? The answer to this lies in Bergson's conception of the vital impetus as a simple 'virtual', the nature of which is to proceed via dissociation. This virtual 'is' only to the extent that it is actualised, and so has to be seen as inseparable from the movement of its actualisation. Through the 'power' of the virtual, lines of differentiation are created. In the case of evolution, then, there is clearly heterogeneity (divergent lines such as plant and animal), but there is also continuity (something of the impetus of life persists across the divergent lines).

This conception of the evolution of life in terms of a virtual multiplicity is opposed to the idea that we are only ever dealing with the 'actuals' of a numerical multiplicity. If we approach evolution in terms of an actual or spatial multiplicity, then time becomes little more than the process of mechanically bringing about the realisation of pre-existent possibilities. The notion of the virtual is opposed to that of possibility. An application of the notion of possibility must be delimited to closed systems; however, in the case of an open system, such as the evolution of life, the notion of a virtual multiplicity is required in order to bring to light its characteristic features. Bergson's conception of a creative evolution can still be put to work today since, as I shall show, contemporary accounts of the evolution of life privilege 'possibility' and operate, unknowingly, on the level of actual multiplicities. Why is a thinking of evolution that focuses on the realisation of the possible so inadequate?

The simple answer to this is that it deprives evolution of any inventiveness or creativity. If the products of evolution are given in advance, in the form of pre-existent possibles, then the actual process of evolution is being treated as a pure mechanism that simply adds existence to something that already had being in the form of a possible. In effect, there is no difference between the possible and the real since the real is simply an 'image' of the possible, its double as it were. If the real merely resembles the possible then we are providing ourselves with a real that is ready-made (*preformed* as it were) and that comes into existence by a series of successive limitations. In the case of the virtual, however, things could not be more different, for here the process of differentiation does not proceed in terms of resemblance or limitation, but rather in terms of divergent lines that require a process of invention. But there is a further aspect to our construction of the possible and the real, one, as we shall see, that plays a crucial role in Bergson's attempt to expose the operations involved when we think events in terms of space and not time (duration): it is not simply the case that the real comes to resemble or mirror the possible, but rather the other way round (the possible resembles the real). This is because we arrive at our notion of the possible by abstracting from the real once it has been made and then projected backwards.

Unfortunately, I do not have the space here to examine in the detail required the 'being' of the virtual and how this ontology is utilised in the conception of a creative evolution. Instead, my aim is simply to show how the contrast between the two processes of the possible and the real and the virtual and the actual can be put to work to cast light on errors and illusions of thought that continue to persist in contemporary accounts of evolution. In demonstrating these errors, we will encounter the criticism made of Bergson by Bachelard, and show how it too falls prey to these same errors.

Despite the characteristically metaphorical comparison of life to an impetus, Bergson's articulation of the notion of 'creative evolution' presents a crucially important philosophical challenge, one that still merits posing today:[10] to what extent can we produce a coherent and consistent conception of *evolution* if we construe it solely and strictly in terms of a set or series of discrete mechanisms (including discrete informational units), ones, it is alleged, that will automatically produce successful adaptations solely through the exogenous workings of natural selection (let us note that it is 'selection' that is doing all the work of 'finality' in the theory)? Can a thinking of 'evolution' be sustained on the basis of this privileging of actual or discrete multiplicities? Some key points are worth stressing at the outset:

1. The claim is not that the scientist has no right to deal with closed systems. Bergson's concern is with what happens when this focus on closed systems, systems from which duration has been artificially extracted, is extended to an explanation of 'life'. His contention is that the focus on closed systems is itself the result of certain 'intellectual tendencies' that have become dominant in the history of our evolution, leading to the ironic result that the human intellect, on account of its spatial habits – which are highly useful for manipulating and regulating matter – is unable to understand adequately its own conditions of existence, that is, unable to comprehend its own creative evolution. Bergson does not deny that there are closed systems. Rather he wishes to point out that isolable systems that can be treated geometrically are the result of a certain tendency of matter, but that the tendency as such is *never fully actualised, never reaches a point of completion*. Indeed, as we shall repeatedly have cause to confirm, the openness of a tendency plays a fundamental role in Bergson's understanding of creative evolution as a continuous movement and of life as virtuality. If science does isolate a system completely for convenience of study, it must still be recognised that a so-called isolated system remains subject to external influences.

2. It is necessary to distinguish between artificial and natural systems, or between the dead and the living. In the case of the living body of an organism the present moment cannot be explained by a preceding moment, as is the case in classical dynamics of matter, since the whole past of an organism needs referring to. An 'artificial' system is one in which time is reduced to a series of discrete 'instants'. But the idea of the immediately preceding instant is a fiction and an abstraction. In effect it denotes that which is connected with a present instant by the interval dt: 'All that you mean to say is that the present state of the system is defined by the equations into which differential coefficients enter, such as $de\backslash dt$, $dv\backslash dt$, that is to say, *present* velocities and *present* accelerations' [CE 22/Œ 512]. In such systems we are only ever dealing with an instantaneous present, one that carries with it a 'tendency', but which it treats as a number (for Bergson, a

Duration and Evolution

tendency has number *only potentially*): '*In short, the world the mathematican deals with is a world that dies and is reborn at every instant – the world which Descartes was thinking of when he spoke of a continued creation.*' [CE 22/Œ 513]

3. While there is a role for calculation and computability (aspects of the present can be calculable as functions of the past), for instance in the realm of organic destruction, this cannot be extended uncritically to all domains, such as organic creation and other evolutionary phenomena that elude mathematical treatment.
4. A coherent conception of 'evolution' requires the notion of duration in which there is a real persistence and prolongation of the past in the present. In a 'natural' system the interval denotes a concrete duration and not simply an extremity. However, duration is implicated in original situations. The novelty of evolution – the 'events' of evolution, if one likes – is to be explained in terms of an indivisible unity, albeit one that admits of differentiation, between 'organic memory' and new conditions or situations (this is in contrast to the research paradigm of neo-Darwinism which conceives evolution taking place in terms of the mechanical sum of discrete genetic codes and the algorithmic process of natural selection). For Bergson the variation of evolution is being produced continuously and insensibly at every moment, although, of course, it is only within specific conditions and under specific circumstances that it gives rise to new species. No amount of knowledge of elementary causes will suffice to foretell the evolution of a new life form.
5. Contrary to widespread misconception which has persisted from Bachelard onwards, Bergson's thinking of 'creative evolution' places a notion of 'radical contingency' at the centre of its concerns and conceives duration precisely in terms of an interruption and discontinuity: duration involves 'incommensurability between what goes before and what follows...' [CE 29/Œ 518–19]. Indeed, it is only by thinking time as duration that the features of rupture and discontinuity can be rendered intelligible.

I

Thinking time as duration is, it is fair to say, the distinctive feature of Bergson's philosophy and informs his conception of metaphysics and critique of science. Science deals with isolable systems and in order to diagram these it eliminates duration. The function of science is to compose a world in which, for the convenience of action, effects of time can be ignored. Although modern physics differs from ancient physics in that 'it considers any moment of time whatever', it still rests on the 'substitution of time-length for time-invention' [CE 342/Œ 784]. Our language supports this elimination since duration is 'always expressed in terms of extension'. In this respect the 'unconscious metaphysics' that sustains science simply conforms to the habits of language which have been fashioned by common sense, and that conform to the actual conditions of life (utility, adaptation, etc.). We shall, of course, have to refine this critique of science simply because it does not characterize the whole of Bergson's take on science. As the remarks about Galileo cited in the first epigraph to this essay show, Bergson is not committed to some *a priori* negative limitation of science. In addition, it is vital that we appreciate the extent to which science can also aid the project of going beyond the human condition. Bergson himself acknowledges this at key turns in his texts. The idea that science is in some *a priori* fashion bound up with a strictly utilitarian and calculative logic is untenable. Although it cannot be demonstrated here, I would suggest that when Bergson limits science to the field of representation he is placing a restrictive empiricism upon its praxis. This is evident, for example, in his critical reception of Relativity when he argues that the multiple times of the theory are *phantasmatic*, that the theory does not lend itself to empirical verification, and that the only 'real' time is that which we can actually perceive and live. My contention is that this reduction of 'conceived time' to 'perceived' and 'lived' time is completely inadequate as a response to the challenge of science and, furthermore, sits uncomfortably with Bergson's own insistence that we learn to think beyond the limits of natural perception ('The insufficiency of our faculties of perception…is what has given birth to philosophy.' [CM 132/Œ 1368]).[11] Moreover, and as we shall see, in his conception of a 'creative evolution' Bergson relies heavily on notions that have no straightforward empirical character (evident in the claim that evolution is characterised by 'tendencies'). Having conceded some ground to the praxis of science, it still needs to be acknowledged that Bergson's critical point that science relies on an unconscious metaphysics, chiefly the metaphysics of a mathematical mechanism, has a certain range of application and validity. This is certainly the case, as we shall shortly see, in contemporary thinking about evolution. Mechanism is not wholly illegitimate or simply false in Bergson's view. Rather, his point is to stress that it is a reflection of our evolved habits of representation rather than an adequate reflection of nature itself. These are habits that

Duration and Evolution 151

conform in large measure to certain tendencies of matter. Mechanism gives us only a partial view of reality and neglects other crucial aspects, such as duration. Hence the need for a philosophy of life.

Bergson situates the impulse towards utility in the context of a specific understanding of human intelligence. Intelligence demands the masking of duration. The intellect is the product of a natural evolution and has evolved as an instrument of praxis or action. Action exerts itself on fixed points. Intelligence, for example, does not consider 'transition', but prefers instead to conceive movement as a movement through space, as a series of positions in which one point is reached, followed by another, and so on. Even if something happens between the points the understanding intercalates new positions, an act that can go on *ad infinitum*. As a result of this reduction of movement to points in space, duration gets broken up into distinct moments that correspond to each of the positions (this is what we can call a discrete or actual multiplicity). Bergson wants to show that these moments of time and positions are only 'snapshots' that the human intellect has extracted from the heterogeneous continuity of movement and the virtual multiplicity of duration (this kind of multiplicity is called continuous or virtual because there is interpenetration between the elements). This is the way in which the mind is able to construct sets and constitute artificially closed systems to aid its diagrammatic designs upon reality. Succession is not the issue, simply because our mind conceives succession in terms of immobilities that are strung out on a segmented line of past, present, and future. As Bergson puts it:

> In short, the time that is envisaged is little more than an ideal space where it is supposed that all past, present, and future events can be set out along a line, and in addition, as something which prevents them from appearing in a single perception: the unrolling in duration (*le déroulement en durée*) would be this very incompletion (*inachèvement*), the addition of a negative quantity. Such, consciously or unconsciously, is the thought of most philosophers, in conformity with the exigencies of the understanding, the necessities of language and the symbolism of science. *Not one of them has sought positive attributes in time.*[12]

The difference to be thought is between an 'evolution', in which continuous phases interpenetrate, and an 'unfurling' in which distinct parts are juxtaposed with each other. In the former case rhythm and tempo are constitutive of the kind of movement in play, so that a retardation or an acceleration are internal modifications in which content and duration are one and the same thing. Throughout his writings Bergson is insistent that states of consciousness and material systems can both be treated in this way. If we say that time merely 'glides over' these systems then we are speaking of simple systems that have been constituted as such only artificially through the operations of our own intellect. Such systems can be calculated ahead of time since they are being posited as existing prior to their realisation in the form of possibles (when a possible is realised it simply gets 'existence' added to it, its fundamental nature has not changed). The successive states of this kind of system can be conceived as moving at any speed, rather like the unrolling of a film: it does not matter at what speed the shots run, an 'evolution' is not being depicted. The reality here is more complex, however, but the complexity is concealed. An unrolling film, for example, remains attached to consciousness that has its own duration and which regulates its movement. If we pay attention to any closed system, such as a glass of sugared water where one has to wait for the sugar to dissolve, we discover that when we cut out from the universe systems for which time is an abstraction, a relation, or a number, the universe itself continues to evolve as an open system:[13]

> If we could grasp it in its entirety, inorganic but interwoven with organic beings, we should see it ceaselessly taking on forms as new, as original, as unforeseeable as our states of consciousness.[14]

One of the difficulties we have in accepting this conception of duration as the invention of the new is due to the way in which we think of evolution as the domain of the realisation of the possible. We have difficulty in thinking that an event – whether a work of art or a work of nature – could have taken place unless it were not already capable of happening. For something to become it must have been possible all along (a conception of logical – spatial – possibility). As Bergson points out, the word 'possibility' can signify at least two different things and often we waver between the two senses. From the negative sense of the word, such as pointing out that there was no known insurmountable obstacle to an event or a thing's coming into being, we pass quickly onto the positive sense of it, in which we hold that any event could have been foreseen in advance of its happening by a mind with adequate information. In the form of an idea this is to suppose that an event was pre-existent to its actual realisation. Even if it is argued that an event, such as the composition of a symphony or a painting, is not conceived in advance, the prejudice still holds sway that such an event *could have been*, and this is to suppose that there exists a transcendent realm of pre-existing possibles.

This reduction of the real, and of real complexity, to mathematical calculability or computation – the very type of reduction which, as we shall see, Daniel Dennett seeks to perform on evolution – is one that Bergson locates in both nineteenth-century physics *and* biology. He quotes the following passage from Du Bois-Reymond's *Über die Grenzen des Naturerkennens* of 1892: 'We can imagine the knowledge of nature arrived at a point where the universal process of the world might be represented by a single mathematical formula, by one immense system of differential equations, from which could be deduced, for each moment, the position, direction, and velocity of every atom of the world' [CE 38/Œ 527]. And this longer citation, highly pertinent to our concerns, from a work of T. H. Huxley's:

> If the fundamental proposition of evolution is true, that the entire world of the living and not living is the result of mutual interaction, according to definite laws, of the forces possessed by the molecules of which the primitive nebulosity of the universe was composed, it is no less than that the existing world lay, potentially, in the cosmic vapour, and that a sufficient intellect could, from a knowledge of the properties of the molecules of that vapour, have predicted, say the state of the Fauna of Great Britain in 1869, with as much certainty as one can say what will happen to the vapour of the breath on a cold winter's day. [CE 38/Œ 527].

Bergson seeks to expose the error of this way of thinking an event or a creative evolution in his essay 'The Possible and the Real'. The error he investigates and exposes is not, as we shall see, restricted to nineteenth-century articulations but also lies at the heart of much current thinking about evolution. Once I have examined further how Bergson addresses the key issues at stake, I shall proceed to show how this same error influences the way Dennett has recently construed evolution by natural selection. No attempt will be made to discredit or refute the thesis of natural selection; rather, we are concerned solely with how an account like Dennett's exemplifies the extent to which any mechanistic approach to evolution produces a spatialised conception of it and fails to account for the continuity of life.

In the essay 'The Possible and Real', Bergson argues that whether we are thinking of the unrolling of our inner lives or that of the universe as an open whole, in both cases we are dealing with 'the continuous creation of unforeseeable novelty'.[15] The first obstacle we have to overcome is that which would posit an opposition between matter and consciousness, or the inert and the living, in which repetition is attributed to the first aspect of the pair, while the qualities of being original and unique are only attributed to the latter. Bergson points out that our focus on the 'inert' is only an abstraction, often serving the need to calculate and map what is solid and simple (matter does have a tendency to inertness but, as a tendency, it is never fully completed).[16] The life of consciousness – growth, ageing, in short, duration – is by no means the preserve of animal life, but can be identified in vegetable life. Moreover, the repetitions of the inorganic world constitute the *rhythms* of creative conscious life and measure their duration. It is owing to the fact that time is a 'searching' and a 'hesitation' that there is a creative evolution in any event or becoming.

This conception of duration as involving a 'searching' and a 'hesitation' is part of how Bergson conceives the virtual as involving an intensive plurality of interpenetrating terms. Only once these terms become extended does it become possible to construct isolable systems that conform to the actual tendencies of matter and within which the only relations are ones of juxtaposition and exteriority. The 'searching' and 'hesitation' do not name anything substantive. Rather, they categorise the general directions, movements, and tendencies of life that are always inventive but in which the products of invention are never given in advance. The notion that there is a 'searching' process of duration within evolution is not peculiar to Bergson. It has most recently been articulated by Manfred Eigen, Nobel prize-winning scientist and Darwinian, in his *Steps Towards Life: A Perspective on Evolution*. Eigen argues that selection 'does not work blindly, and neither is it the blind sieve that, since Darwin, it has been assumed to be.'[17] The problem with Eigen's account, however, is that it treats the active searching feedback mechanism of selection solely in terms of an over-arching tendency that is always actual – with natural selection one always knows what is going to be produced in advance as a general law of evolution (survival of the fittest and successful adaptations, a thesis that has frequently been recognised as tautologous). The key aspect of time for Bergson is that it introduces *indetermination* into the very essence of life. This indetermination does become materially embodied: a nervous system, for example, can be regarded as a 'veritable reservoir of indetermination' in that its neurons 'open up multiple paths for responding to manifold questions' posed by an enviroment [CE 125/Œ 602]. However, our natural or instinctive bent is always to construe this indetermination in terms of a *completion* of pre-existent possibilities. This is because we are subject to habits of thinking that have evolved from conditions of evolutionary life, which are conditions of utility and adaptation:

Perception seizes upon the infinitely repeated shocks which are light or heat...and contracts them into relatively invariable sensations: trillions of external vibrations are what the vision of a colour condenses in our eyes in the fraction of a second... To form a general idea is to abstract from varied and changing things a common aspect which does not change or at least offers an invariable hold to our action. The invariability of our attitude, the identity of our eventual or virtual reaction to the multiplicity and variability of the objects represented is what first marks and delineates the generality of the idea.[18]

The intellect, which has evolved as an organ of utility, has a need for stability and reliability. It thus seeks connections and establishes stable and regular relations between transitory facts. It also develops laws to map these connections and regularities. This operation is held to be more perfect the more the law in question becomes mathematical. From this disposition of the intellect emerges the specific conceptions of matter that have characterised a great deal of Western metaphysics and science. Intelligence, for example, conceives the origin and evolution of the universe as an arrangement and rearrangement of parts which simply shift from one place to another. This is what Bergson calls the Laplacean dogma that has informed a great deal of modern enquiry, leading to a determinism and a mechanism in which by positing a definite number of stable elements all possible combinations can be deduced without regard for the reality of duration [CE 38/Œ 527].

Most of the anxieties of metaphysics concern problems that have been badly posed. In 'The Possible and the Real' Bergson reduces many of these problems to the way in which we construe a negative as containing 'less' than its positive opposite, such as is found in the pairs 'nothing/being', 'disorder/order', and 'possible/real'. These evaluations, however, reflect the habits of our intellect. Bergson ingeniously shows that there is, in fact, more intellectual content contained in the negative ideas than in the positive ones. This is because they draw on several orders and several existences in order to make themselves intelligible. For example, what does saying that the universe is 'disordered' exactly mean? What is the status of the appellation 'disorder' in this case? Here the idea of disorder is posited on the basis of a conception of order we already have, which leads us to positing this disorder in relation to an order we expect to find but do not. In the intriguing case of the 'possible' our guiding habit is to suppose that there is less contained in the possible than in what is actually real and, consequently, we hold that the existence of things is preceded by their possibility. Bergson once again shows with this example that the reverse is in fact the case. If we take the example of living things 'we find that there is more and not less in the possibility of each successive state than in their reality.' This is because the possible only precedes the real through an intellectual act that conceals its own illusion with regard to the issue. Bergson expresses the key insight, puzzling on first encounter, as follows, '...the possible is only the real with the addition of an act of mind which throws its image back into the past, once it has been enacted.'[19] What does this mean? If we accept that reality is implicated in duration, and that this duration involves the creation of the new in the sense of something that is unforeseeable and incalculable, then we arrive at the insight that we ought not to say that the possible precedes its own reality but rather than it *will have preceded it once the reality has appeared*. In short, the new is able to create its own conditions of possibility. It is owing to the fact that we break up the passage of time into discrete segments and isolable instants or phases that we are led to positing an unrolling in terms of the realisation of possibles. An event happens, a work of nature or a work of art is created or comes into being, and we then construe its possibility in terms of a mirage of the present in the past. Owing to our construction of time as a linear succession of stages and instants we know in advance that every future will ultimately constitute a present, every present will become a past, and so on. It is by way of such illusions that we regulate our individual and social lives. The illusion is real, and what Bergson is concerned with is bringing out the genuinely creative aspects of the living that it misses and overlooks. Of course, he is not denying that there is an arrow or movement of time from past to future; his point is rather that time involves a living interpenetration rather than a juxtaposition of discrete phases or stages. If the past is not co-existent with the present then how can the present pass?[20]

So, the possible is only posited from the vantage point of the real or when something has become actual. In other words, it works in a contrary fashion to the way we habitually suppose. Of course, Bergson does not deny that we can construct closed systems in which the relation between the possible and the real would conform to our intellectual expectations. But this is because such systems have been artificially carved out, made subject to the regularity of mathematical or physical laws, and rendered isolable because duration has been left outside the system. If, however, we understand the illusion of the possible and show how it is generated, we come to appreciate that evolution concerns something quite different from the realisation of a programme. The point is an important one for Bergson since it reveals to him that the 'gates of the future' are open. There is indetermination in life owing to the fact that the universe is an open system. This indetermination is not to be confused with a competition between pre-existent possibles.

Bergson accuses philosophy of a great refusal in the face of the reality of time and its creation of unforeseeable novelty. We are all born Platonists to the extent that we imagine that Being is given once and for all in some immutable system of Ideas.[21] On this model time can only appear as degeneration and distortion of real essence. Variation and difference are tantamount to deficiencies. Surely, however, we are now all born Darwinians, philosophers and scientists of variation and difference, fully cognisant of the fact that we are the creatures of a *natural* selection involving millions and millions of years of geological time, and that the whole of nature is subject to processes of growth, ageing, decay, degeneration, transformation, and so on? There remains within Darwinism, however, an inability to think evolution as duration. This is a major deficiency in the theory since it means that Darwinism remains wedded to a thinking of life in terms of the realisation of the possible. This, as I shall show, is a direct effect of approaching evolution in terms of a mechanistic and algorithmic process. As Bergson says, Darwinism is well able to explain the sinuosities of the movement of evolution, but as for the *movement itself*, it has no conception; that is, it correctly describes the trajectory and actual history of evolution, but is unable to account for the continuity of evolution as a process of 'life'. In short, Darwinism remains on the level of actual or discrete multiplicities. It cannot think life in terms of the 'virtual' (or in the infinitive, as Adamson has put it).[22] A review of Dennett's presentation of Darwin's 'dangerous idea' will exemplify these points.

II

Bergson is not an anti-Darwinian thinker. His own thinking is neither possible nor intelligible – including his understanding of the human condition, a condition that is implicated in the *conditions of evolutionary life*, and the desire to think beyond it – without an appreciation of the import of Darwinism. The doctrine of 'transformism' (as Bergson calls natural selection) now forces itself, he says, upon all philosophy and science, making it impossible to speak any longer of 'life' as an abstraction. 'Life' is a continuity of genetic energy that cuts across the bodies 'it has organized one after another, passing from generation to generation, [and that] has become divided among species and distributed amongst individuals without losing anything of its force, rather intensifying in proportion to its advance' [CE 26/Œ 516]. Bergson is, in fact, open to the different accounts of evolutionism that modern thought provides. He holds, for example, that the whole issue of the transmission of acquired characteristics cannot be settled either by making an appeal to vague generalities about evolution or by closing it down through some *a priori* conception of the nature of evolution and of what is and is not possible; rather, it is strictly a matter of further open empirical inquiry and experimentation [CE 78/Œ 562].

Bergson does not go on to embrace a finalist position. He argues that finalism is merely an 'inverted mechanism' which also reduces time to a process of realisation. In the doctrine of teleology, for example, evolution is construed as the realisation of a programme previously arranged and ordered. Again succession and movement remain mere appearances, and the attraction of the future is substituted for the impulsion of the past (hence the inversion). In Leibniz time is reduced to a confused perception that is entirely relative to the human standpoint. For a mind seated at the centre of things there would be no time and the confused perception would vanish. The only notion of finality Bergson will permit, contra Leibniz and Kant, is a strictly *external* finality.[23] However, while conceding that actual change is something accidental he insists that the *tendency to change* is not [CE 85/Œ 568]. If there were no such tendency it is difficult to see how the idea of an 'evolution' could be made intelligible. In short, Bergson argues that the evolution of life cannot be treated either simply in terms of adaptations to accidental circumstances or as the realisation of a plan or programme.

In Bergson's conception of creative evolution the notion of 'tendency' serves an important function. He first introduces the word in the opening argument of the book where it denotes the 'directions' of life and serves to counter the idea that there is a single universal biological law that can be applied automatically to every living thing, and the assumption that evolution can ever be made up of *completed* realities: it is thus a movement which remains *open*. For Bergson, evolution is marked by different and conflictual tendencies; for example, life reveals a tendency towards individuation and a tendency towards reproduction. The notion of tendency is also designed to suggest that the study of life can be approached in terms of problems that are *immanent* to an evolutionary process or movement. The directionality and movement of life are not, however, to be understood in terms of a simple mechanical realisation of pre-existing goals. Rather, the 'problems' of evolutionary life are general

Duration and Evolution

ones, denoting an immanent virtual field that is responded to in terms of specific 'solutions' (an example to illustrate this would be cases of homology within evolution, such as the eye, representing solutions to general problems that are common to different phylogenetic lineages, in this case that of light and the tendency 'to see', or vision). Bergson is struck by the fact that evolution has taken place in terms of a *dissociation* of tendencies and through divergent lines that have not ceased to radiate new paths. Now, his speculative claim is that this dissociation of tendencies and divergency of lines can best be explained in terms of an indeterminate and inchoate initial impetus (a simple 'virtual'). The evolution of life, as an evolution, becomes intelligible when it is viewed in terms of the *continuation* of this impetus that has split up into *divergent* lines. So we have in this thinking of creative evolution *both* continuity and discontinuity. Let us also note that there is on Bergson's model no *dominant tendency* within evolution and neither can the different forms of life be construed in terms of the development *of one and the same tendency*.

Bergson holds that his way of thinking can best account for cases of convergent evolution that are at the heart of the inquiries of Darwinism. The claim is that this initial impulsion of life continues to abide in the 'parts', so explaining the evolution of identical organs in very different forms of life. It is owing to this common impetus that it is possible to explain the different solutions divergent lines of evolution, such as plant and animal, come up with in response to problems of storing and transforming energy: 'the same impetus that has led the animal to give itself nerves and nerve centres must have ended, in the plant, in the chlorophyillian function' [CE 114/Œ 592]. Bergson is contesting the claim that cases of convergent evolution can be explained simply in mechanistic terms as the mere accumulation of a discrete series of accidents added to one another and preserved through selection. Of course, he is aware of the argument which would insist that resemblances of structure across very different organisms are the result of a similarity in the general conditions under which life has evolved. But the weakness of this argument is that it suggests that external conditions alone are sufficient to bring about a precise adjustment of an organism to its environmental circumstances (and his claim is that in natural selection 'adaptation' is equivalent to 'mechanical adjustment'). Hence his question and problem: 'How can accidental causes, occurring in an accidental order, be supposed to have repeatedly come to the same result, the same causes being infinitely numerous and the effect infinitely complicated?' [CE 56/Œ 543]. As he points out, that two walkers commencing from different points and wandering at random should finally meet is not a great wonder. However, what is surprising is that throughout their perambulations both walkers should describe two identical curves that are superposable upon each other. Furthermore, Bergson is more than willing to concede that the first rudiments of the eye can be found in the pigment-spot of lower organisms and that this was probably produced purely physically by the mere action of light, and that between this simple pigment and the complicated eye of a vertebrate there are a great number of intermediaries. But, as he then points out, 'from the fact that we pass from one thing to another by degrees, it does not follow that the two things are of the same nature' [CE 70/Œ 555]. In short, what is missing in the mechanistic conception of adaptation is any sense that in certain life forms the evolution of organs cannot simply be explained in terms of the passive adaptation of inert matter submitting to the influence of an environment. The simple influence of light cannot be held to be the cause of the formation of the various systems (nervous, muscular, osseous) that are continuous with the apparatus of vision in vertebrate organisms [CE 71/Œ 556]. But in addition to this he is also claiming that across different lineages the same virtual problem has persisted. The attempt to account for convergency in terms of an initial impulsion of life is clearly the most speculative aspect of Bergson's thinking about evolution.

His general aim can, however, be upheld, which is not one of simply attacking mechanism but rather asking serious questions about the precise character of the mechanisms of life and the nature of adaptation. Bergson often argues that the real issue at stake is precisely what *kind* of mechanism we have in mind.[24] Dennett insists on approaching natural selection as an entirely mechanistic process based on algorithmic designs. Natural selection aims to show how a non-intelligent – that is, robotic and mindless – artificer is able over periods of time to produce successful adaptations. According to Dennett, Darwin's celebrated 'one long argument' is composed of two demonstrations, a logical one which claims that a 'certain *sort* of process would necessarily have a certain outcome' and an 'empirical' one that aims to show that the 'requisite conditions' for such a process can, in fact, be identified in nature.[25] The two demonstrations come together when it is shown, Dennett claims, that at the heart of Darwin's discovery is the power of an algorithm. This he defines as 'a certain sort of formal process that can be counted on – logically – to yield a certain sort of result whenever it is "run" or instantiated' [DDI 50]. In other words, evolution is a programme and in its actualisations it simply *instantiates*. It has a design, albeit that of a

Duration and Evolution 161

nonintelligent and mindless artificer, which is able to be instantiated to produce certain results whenever it is programmed to 'run'. An algorithmic 'process' has several features, but one of the most salient ones is that it is made up of constituent steps. These are mindless steps which produce 'brilliant results'. Any 'dutiful idiot' or straightforward mechanical device could perform these to make the machine of selection yield the necessary results (successful adaptations).

Now, Dennett does *not* want to claim that results generated within evolution are conceived in advance. He writes, for example, 'Evolution is not a process that was designed to produce us, but it does not follow from this that evolution is not an algorithmic process that has in fact produced us' [DDI 56]. Conceived as an algorithmic process natural selection is not about what it will inevitably produce but what it may or is most likely to produce and what it will 'tend' to yield.

> Here, then, is Darwin's dangerous idea: the algorithmic level *is* the level that best accounts for the speed of the antelope, the wing of the eagle, the shape of the orchid, the diversity of species, and all the other occasions for wonder in the world of nature. It is hard to believe that something as mindless and mechanical as an algorithm could produce such wonderful things. No matter how impressive the products of an algorithm, the underlying process always consists of nothing but a set of individually mindless steps succeeding each other without the help of any intelligent supervision; they are 'automatic' by definition: the workings of an automaton. [DDI 59]

The object of Dennett's attack is, of course, any non-naturalistic account of evolution. However, Bergson's quarrel with Darwinism is not over an anti-naturalism versus a naturalism, in which a mindless mechanism would be replaced with something mindful. The much lambasted but ill-understood notion of *élan vital* is bound up with immanent *tendencies* of evolution, not transcendent to evolution, and concerns the play between certain entropic tendencies of matter and certain creative tendencies of 'life' (individuation, reproduction, and survival itself). Bergson's conception of a creative evolution is, in fact, working against the idea of there being some transcendent mind – which he thinks is modelled on the habits of our intellect anyhow – that would be able to design evolution in advance. Evolution for him 'remains creative' even in its adaptations. In other words, natural selection, as one key component in evolution, is a creative not a mechanical process. The fact remains, however, that Dennett can only think evolution in terms of logical possibility. For Dennett, Darwinism is all about evolution by design, and it just happens that this takes place via mindless, mechanistic means.

Throughout his text Dennett persists in positing spurious oppositions, a key one being that between a 'crane' and a 'skyhook', in which the former refers to the discrete stages that characterise an algorithmic process, while the latter corresponds to the failure of nerve that characterises any approach that has recourse to some kind of *deus ex machina* to explain the origins and development of something. But this limits the routes available to theoretical and empirical enquiry far too narrowly. The choice is far too simple-minded: *either* the algorithms of natural selection *or* some appeal to divine or special creation. In Dennett's terms everything from Nietzsche's 'will to power' to Bergson's *élan vital* would be readily dismissed as conforming to the latter strategy. The problem here is that he has no serious philosophical appreciation of why these thinkers felt compelled to introduce such notions into their accounts of evolution (in Nietzsche's case matters are complicated since the 'will to power' functions as both a transcendental and an empirical principle; the same could, of course, be said of Bergson's *élan vital*).[26] Moreover, the crane/skyhook couplet simply leaves untouched some crucial questions about the nature of an evolution, and these are at the heart of Bergson's concerns (such as, to give one example, the nature of matter's self-organising capacities).

Let me now show how Dennett is restricted in his conception of evolution on account of his attachment to thinking it largely in terms of the possible. Indeed, there is a whole chapter in his book, the title of which takes its inspiration from a work by François Jacob, called 'The Possible and the Actual'. At the heart of his conception is the idea of a 'design space'. Such an idea corresponds exactly to Bergson's exposition of how the intellect – an organ for manipulating matter and manufacturing tools and implements – reduces time to space. Dennett claims: 'There is a single Design Space in which the processes of both biological and human creativity make their tracks, using similar methods' [DDI 123].

The famous Tree of Life is to be thought in terms of this design space, meaning that the actual trajectories and tracks of evolution are to be understood as 'zigzagging' through a vast multidimensional space, 'branching and blooming with virtually unimaginable fecundity', while managing only to fill 'a Vanishingly small portion of that space of the Possible with Actual Designs' [DDI 143]. What is meant exactly here by the 'possible' and the 'actual'? This comes out most clearly in the idea Dennett develops of a 'library of Mendel', which he offers as a variant of

Duration and Evolution 163

Borges' library of Babel and which he develops as a way of answering 'difficult questions about the scope of biological possibility'. In Borges' library there lies a potential infinity of possible books that could be written. Not only do we find *Moby Dick* in there but also a million 'impostors', each one of which differs from the real one by a single typographical error. All these books, and billions more besides, exist in the library of some virtual or possible but stupendously vast logical space. The problem is how to search, locate, and find the book we might want to actualise, such as the biography of one's life (which may exist in multiple forms). The 'Library of Mendel' is constructed as a biological variant of this logical space of all possible books, containing all possible genomes or DNA sequences. Possible genomes and sequences here refers, of course, to what we know of life on planet Earth, so just as Borges' library ignored books composed of other alphabets (Chinese, for example), so the library of Mendel excludes genetic codes not yet known to us. Of course, the analogy between the two libraries does not strictly hold. Dennett, for example, reflects on the chemical stability of his library, noting that all the permutations of the sequences of DNA (adenine, cytosine, thymine, and guanine) enjoy this stability and that all could conceivably be constructed in principle in a gene-splicing laboratory. But then he notes that not every sequence in this library corresponds to a 'viable organism', simply because many if not most DNA sequences are 'gibberish' – 'recipes for nothing living at all' [DDI 113]. Of course, in the library of Babel one can well imagine that the vast majority of books would be of this type – the gibberish of *Finnegan's Wake*, for example. Of course, we need to ask, not only whether this is an 'original' book or a mere impostor, but also, to complicate things dramatically, by considering whether the so-called 'original' text might not itself be the impostor. Such examples could be readily multiplied but none of these books would for this reason be considered a lesser book – as in a failed adaptation – as a result of being composed of 'gibberish'. Indeed, what this example shows is that natural selection fails to acknowledge its own reliance on an over-arching *tendency* to explain the directions of evolution, one that the theory runs the risk of positing as a substantive transcendent principle, namely, survival of the fittest. But how can it account for this 'tendency' as the overriding tendency of evolution? If the entire burden of proof is placed on natural selection does this not run the risk of turning the mechanism of 'selection' itself into a *deus ex machina*?

All the time that Dennett thinks about logical genetic possibility and actual evolution he is thinking spatially. This explains why, for example, he is able to hold to the position that tigers were, all along, a logical possibility in this design space of the library of Mendel. But what is the logical sense of this view? In truth, Dennett's speculation on the possibility of tigers conforms precisely to Bergson's insight that the construction of the possible takes place only in terms of retrospection. Dennett, for example, writes: 'With hindsight, we can say that tigers were in fact possible all along, if distant and extremely improbable' [DDI 119]. But could we not say this, with the benefit of hindsight, of anything that now exists (that it was in fact possible all along if somewhat improbable)? What is the *empirical* weight of such a claim? Dennett's insistence that he is concerned not with what is possible in principle in this library of Mendel but with what is 'practically possible' makes no difference to the force of our objection to his construal of a mechanical evolution. The problem with it is that it allows no room for thinking evolution in terms of duration, and to think this duration in relation to the real complexity of a component system (genes, organism, environment, etc.). Dennett's argument is, in fact, deeply metaphysical and empirically worthless. As G. Adamson points out in a recent wide-ranging article on Bergson and evolution, Dennett's conceptions of probability and possibility 'hide some fairly extravagant epistemological assumptions.' However, it is not that Dennett denies 'Darwin's basic intuition that forms are created', as Adamson claims, but that he construes such creativity in terms of the instantiation of an algorithmic 'process'.[27]

If any further proof were needed that Dennett reduces the time of evolution to the space of logical and genetic possibility, consider his treatment of why certain non-actual possibles didn't in fact happen, such as all your non-actual brothers and sisters. The answer he gives is the intelligibly straightforward one that your parents didn't have the time, desire, or energy, and ultimately no reason can be given for this unfortunate or fortunate state of affairs. But read carefully the way he argues this:

> As the actual genomes that *did* happen to happen began to move away from the locations in Design Space of near misses, their probability of ever happening grew smaller. They were so close to becoming actual, and then their moment passed! Will they get another chance? It is possible, but Vastly improbable, given the Vast size of the space in which they reside. [DDI 125]

One wants to ask, not only what is the 'space' being spoken of in this passage, but also: what does it mean to posit 'near misses' that never came into existence and that probably never will, and to talk of things being close

Duration and Evolution

to 'becoming actual'? How are these conceptions actually constructed by our intellect? For surely what Dennett is saying in this bizarre passage reveals far more about the nature of the intellect than it does about the actual nature of evolution. I'd like to quote a fairly long passage from Bergson, which contains genuine insight coupled with his customary clarity and lucidity. The passage captures both the confused working of the logic of the possible, to which, as the preceding passage makes clear, Dennett is committed; but also precisely how Bergson's notion of a creative evolution enables a genuine thinking of novelty:

> Our ordinary logic is a logic of retrospection. It cannot help throwing present realities, reduced to possibilities or virtualities, back into the past, so that what is compounded now must, in its eyes, always have been so. It does not admit that a simple state can, in remaining what it is, become a compound state solely because evolution will have created new viewpoints from which to consider it... Our logic will not believe that if these elements had sprung forth as realities they would not have existed before that as possibilities, the possibility of a thing never being (except where that thing is a purely mechanical arrangement of pre-existing elements) more than the mirage, in that indefinite past, of reality that has come into being. If this logic we are accustomed to pushes the reality that springs forth in the present back into the past in the form of a possible, it is precisely because it will not admit that anything does spring up, that something is created and that time is efficacious. It sees in a new form or quality only a rearrangement of the old and nothing absolutely new. For it, all multiplicity resolves itself into a definite number of unities. It does not accept the idea of an indistinct and even undivided multiplicity, purely intensive or qualitative, which, while remaining what it is, will comprise an indefinitely increasing number of elements, as the new points of view for considering it appear in the world. To be sure, it is not a question of giving up that logic or of revolting against it. But we must extend it, make it suppler, adapt it to a duration in which novelty is constantly springing forth and evolution is creative.[28]

Dennett himself raises a key question in this chapter of the book when he asks, is it possible to measure Design, albeit imperfectly? This, he then goes on to say, involves dealing with the question of whether Darwinian mechanisms are powerful and efficient enough to have done all the necessary work in the time required. Again, we see the *a prioristic* and assumptive nature of Dennett's position: evolution is work and has things to do and to produce; time is conceived as the space (a stretch or span of time) in which these tasks are actually done, executed, performed, etc. He himself poses the issue in terms of the error of analysing genomes – and allied phenomena such as random drift, etc. – in isolation from the organisms they create. In order to get any serious purchase on the issue it is necessary, he argues, 'to look at the whole organism, in its environment' [DDI 127]. Now this is a good Bergsonian move to make, since it complicates massively the algorithmic picture we have so far been blackmailed into accepting (if you do not believe in evolution as an algorithmic process we will 'out' you as a closet skyhooker). Of course, Dennett believes he can speak in terms of the vast possible and the finite actual because he has a correct appreciation of the material facts and details of evolution. However, this means that he is wedded to some unempirical ideas about the nature of open systems. It is only artificially that he can put back together the pieces of the jigsaw which he has separated by an act of abstraction (genes, organisms, the environment, time, etc.).

Dennett's Darwinism contains enormous amounts of mystification and reification, which is troubling for a theory that purports to be a radical materialism. His attachment to the power of the algorithm leads him to conflate all kinds of complicated processes of evolution (such as the biological and the cultural) and which also result in him posing problems as serious that are, in truth, risible. According to Dennett, the key lesson to be learned from Darwin's revolution is this: Paley was right in holding Design to be not only a wonderful thing but also to involve 'intelligence'. Darwin's contribution was to show that this 'intelligence' could be broken up into 'bits so tiny and stupid that they didn't count as intelligence at all, and then distributed through space and time in a gigantic, connected network of algorithmic process' [DDI 133]. He insists that there is only *one* Design space and everything from the biological to the social and technological evolves from it and, moreover, that everything actual in this space is united with everything else [DDI 135)]. As both designed and as designers 'we' ourselves manufacture products in terms of the non-miraculous logical power of the algorithm and always in accordance with the blind and mechanical process of selection. Dennett asks us to reflect on the following 'problem' as a genuinely serious problem:

> How many cranes-on-top-of-cranes does it take to get away from the early design explorations of prokaryotic lineages to the mathematical investigations of Oxford dons? That is the question posed by Darwinian thinking. [DDI 136]

If *this* is the question that lies at the heart of Darwinian thinking, the difficulty is not in determining what is dangerous or radical about it but rather why we should take it seriously. It is not sufficient for Dennett to proclaim that anyone who cannot recognise this as a serious question is someone who cannot, perhaps on account of deeply rooted existential resistances and traumas, come to terms with the fact that they and their consciousness are the product of mindless robots and purely mechanical processes. Positing a Manichean world of craners and skyhookers is a neat but unsubtle way of considering the complex questions at issue and the difficult problems at stake, which are in need of a less simple-minded approach than is on offer in his work. Dennett's exposition of Darwin's dangerous idea is, I believe, an instructive example of the extent to which much contemporary thinking about evolution remains in the grip of spatialised habits.

III

Bergson insists that he does not depart from the fundamental axiom of scientific mechanism, namely, that there is an identity between inert matter and organized matter. Instead he shifts the ground of the question by asking whether the natural systems of living beings are to be assimilated to the artificial systems that science cuts out within inert matter. If there is a mechanism of life this is a mechanism of the 'real whole', and not simply that of parts artificially isolated within this whole [CE 31/Œ 520]. The isolable systems that are cut out of this indivisible continuity are not actually parts at all but rather *'partial views'* of this whole. Chemistry and physics are unable to provide us with the 'key to life' because they simply put these partial views end to end. No reconstruction of the whole is possible on this basis, which would be like multiplying photographs of an object in an infinite number of aspects in a vain effort to reproduce the object. The isolable or closed systems that the intellect carves out from the real are, of course, not mere fictions; rather they correspond to actual tendencies of matter itself. As we shall see, there are two different accounts of the 'whole' in Bergson, what one might call the 'whole' that refers to the initial source of life's impulsive character, and the 'whole' that refers to the ever-changing set of relations which characterise the movement of the systems of life within the universe. What is important with respect to the first conception is that we do not treat it in spatial terms. How and why this is to be avoided, I will come to shortly.

Bergson's argument that movement is irreducible is well known. Movement cannot be reconstituted from either positions in space or instants in time. If it is said that we do this by adding to the positions or instants the idea of a succession, it is not being recognized that this move is equally abstract since it consists of a time that is mechanical and homogeneous, one that has been copied from space and that is valid for all movements. It is his adherence to the irreducible character of movement which informs Bergson's contention that life is not reducible to its physico-chemical basis. Just as we may legitimately ask whether a curve is composed of straight lines, so we can ask whether 'evolution' is made up of discrete stages and isolable systems. For Bergson evolution can be thought in terms of a 'single indivisible history' [CE 37/Œ 526], but this is a whole that is best conceived in terms of a 'virtual Open'. Mechanism errs in focusing attention only on those isolable systems that it has detached from this 'whole'. A mechanical explanation is only possible through such an artificial extraction. It is with this conception of the 'whole' that it is possible to show the limits of the criticism that has been levelled at Bergsonism from Bachelard to Badiou.[29]

Bachelard declares that he accepts 'everything' of Bergsonism except the thesis on continuity. He turns to 'sets' as an example where *discontinuity* is established and which decides the nature of the continuum:

> We do not feel we have the right to impose a continuum when we always and everywhere observe discontinuity; we refuse to postulate the fullness of substance since any one of its characteristics makes its appearance on the dotted line of diversity. Whatever the series of events being studied, we observe that these events are bordered by a time in which nothing happens. You can add together as many series as you like but nothing proves that you will attain the continuum of duration. It is rash to postulate this continuum, especially when one remembers the existence of mathematical sets which, while being discontinuous, have the power of a continuum. Discontinuous sets such as these can in many respects replace one that is continuous.[30]

Some key points need to be noted in response to this passage: firstly it is not that Bergson does not have a notion of discontinuity and that all we have in his system is a plenitude of continuity (it is difficult to see how Bergson's duration would differ from an unchanging substance if this were the case). We say this not simply because Bergson construes the organism in terms of a discontinuity within the flow of genetic energy that characterizes life, but rather that discontinuity – in the form of the dissociation of tendencies and the divergency of lines of evolution – is an integral and essential part of his conception of the continuity of life. It comes to matter greatly, therefore, how we construe the relation between the discontinuous (the parts) and the continuum (the whole).

The potential problem with Bergson's account lies in his description of the original impetus as a virtual whole, and in particular, when he speaks of it as a point of origin or 'source' [CE 54/Œ 541]. This is problematic because it suggests that the whole precedes and exists prior to the actual parts and that it is, in some sense, given. But this cannot be right. It is as if Bergson is asking us to think evolution on two different planes, equally real, at one and the same time: on the plane of a pure virtual in which the tendencies of life have not yet been actualised and so exist in terms of an intensive fold[31] (a monism); and on the plane of an actualised virtual in which there are only divergent lines with forms of life, such as animal and plant, becoming closed on themselves, constituting an unlimited pluralism. But this still does not show how it is possible to think the virtual outside the terms of the 'given'. This is done through two ways, and both are necessary in order to avoid the trap of the given. On the one hand, it is necessary to show that the virtual cannot be separated from a process of actualisation (except, of course, in abstract terms). On the other hand, it is necessary to recognise that the nature of the virtual is constantly changing, and must be so. As Deleuze says: 'Each line of differentiation or actualization thus constitutes a "plane (*plan*) of nature" that takes up again in its own way a virtual section or level.'[32] When Bergson speaks of reintegrating the systems that science isolates into the 'Whole', this must denote a whole that at any point in time is never ceasing to become (it is a pure becoming). As Deleuze writes on the character of this open whole – a whole not amenable to set theory (unless it be constructed as the infinite set, the set of all sets) – '…if the whole is not able to be given, it is because it is the Open, and because its nature is to change constantly, or to give rise to something new, in short, to endure'.[33]

In his reading of *Creative Evolution* in the 1966 text *Bergsonism*, Deleuze insists that, contrary to appearances, the simple virtual whole is *not* given. Rather, it is only when we assimilate time to space that it becomes so. It is only in artificial terms that the whole, as virtual, can be thought in abstraction from its actual divisions and movements (by turning time into space). This is important, since it shows that the actualisation of the virtual is not to be conceived in terms of a logic of negation (the actual forms of life created are not degraded forms of some transcendent and immutable being), but rather solely in terms of acts of positive creation. The virtual, then, is neither a Platonic form or Idea of evolution, nor a supplementary dimension existing in some given realm that would be transcendent to the 'actuals' of evolution. It is only if we conceive it in terms of space that it assumes the appearance of such a dimension. Admittedly, it is difficult to think the virtual in this way, but the difficulty demonstrates, I would suggest, the tremendous efforts required of us to think non-spatially.

In declaring that he accepted everything of Bergsonism except continuity, Bachelard was drawing attention to the need to show that continuities can never be regarded as complete, solid, and constant. Rather, they 'have to be constructed'.[34] This means for Bachelard that the continuity of duration cannot be an immediate datum of consciousness but has to be conceived as a problem. As we have seen, however, this is precisely how Bergson comes to construe duration in *Creative Evolution*, as a problem of parts and the mobile virtual whole, of relatively isolable systems and the threads that connect these systems, including the invisible bonds that maintain a solidarity and a communication between diverse forms of life, to the rest of the universe. Evolution is characterised by tendencies and problems – this is what is meant by a 'biological metaphysics' – in which the new is not possible in an 'instant' of time, but requires the elaboration of an evolutionary time.[35] Where Bachelard goes wrong is in supposing that we have to choose between continuity and discontinuity, that we can only have sets and parts at the expense of the whole and wholes, and that in order to allow for the new we have to sacrifice duration. What he fails to appreciate in Bergson is this innovative way of thinking systems in both their actual and virtual complexity. In arguing for the need to see a 'fundamental heterogeneity at the very heart of lived, active, creative duration',[36] he failed to see that such heterogeneity is already at the heart of a Bergsonian appreciation of duration.

We must confess to finding Bachelard's treatment of Bergson on evolution, evident in the essay on the 'Instant' collected in this volume, superficial. This is most apparent in the naïveté contained in his assertion that 'Bergson no doubt had to ignore accidents when writing his epic account of evolution.'[37] This both disregards the important role Bergson ascribes to contingency within evolution and simply fails to engage with his examination of the inadequacy of Darwinism as a theory of *evolution*.[38] Evolution cannot simply be explicable in terms of a mechanical adjustment to external conditions or circumstances. Bergson argues, for example, that the theory of mechanism cannot explain a crucial element in the evolution of the eye, namely, 'correlation'. On the one hand we have a complex organ, and on the other we have a *unity* and simplicity of function. It is this contrast, says Bergson, which should make us pause for thought. If vision is 'one simple fact' how is it possible to account for its organisation and operation in purely exogenous terms and in terms of chance modifications [CE 88/Œ 570]? If we are to take seriously the idea that a complex organ like the eye was the result of a gradual formation, as well as of a process of highly complex correlation, then it becomes necessary to attribute to organised matter the power of constructing complicated machines able to utilise the excitations that it undergoes [CE 72/Œ 556]. Bergson makes it clear, in responding to a critical point on utility which would argue that the eye is not made to see but that creatures see because they have eyes, that he is not simply referring to an eye that has the capacity to see when speaking of an eye that 'makes use of' light. Rather, he is saying that what needs paying attention to are the precise relations existing between the organ and the apparatus of locomotion. In other words, the problem is not that of a discrete organ, such as the eye, but the complexity of its evolution in relation to other systems of an organism.

Bergson is not tied to the idea that there are no 'accidents' or contingencies in evolution, or that we have to posit evolution in terms of a linear and direct process; on the contrary, for Bergson there is a process characterised by frequent dead ends, numerous aborted lines, and lines that have failed to evolve. On the other hand, however, 'the failures and the deviations of the transformist mechanisms have not arrested the increase in either anatomical or psychic complexity.'[39] Evolution has developed along a plurality of lines. Vitalism exists to remind us of our ignorance regarding important questions of evolution. Mechanism simply invites us to ignore this ignorance. Bergson fully concedes that if the universe as a whole is carrying out a plan, then this is something which can never admit of an empirical demonstration. What we do know is that nature 'sets living beings at discord with one another' and 'everywhere presents disorder alongside of order, retrogression alongside of progress' [CE 40/Œ 529]. One of the reasons why vitalism is such an intangible position to hold is because of its crucial claim, at least as articulated in Bergson, that in nature there exists neither purely internal finality nor absolutely distinct individuality: 'The biologist who proceeds as a geometrician is too ready to take advantage here of our inability to give a precise and general definition of individuality. A perfect definition applies only to a *completed* reality; now, vital properties are never entirely realized, though always on the way to become so; they are not so much *states* as *tendencies*' [CE 12–13/Œ 505].

IV

This is no teaching of immediate intuition.
Bergson, *Matter and Memory*[40]

For Bergson time conceived as duration denotes nothing mysterious, unempirical, or ineffable. On the contrary, it lends itself to clear and precise philosophical analysis once we have learned to think beyond certain mental habits and reifications. Real duration denotes the 'indivisible continuity of change'. It involves movement, it is movement and change, and the conception we impose upon the flux of becoming of a successive series of 'befores' and 'afters' simply reflects the need we have cultivated to impose a uniform and regular order upon it. By confusing intensity with extensity, and succession with simultaneity, we import spatial images into time that enable us to introduce clear-cut distinctions of discrete moments and parts external to one another. Of course, reality is made up of both extensity and duration, but this 'extent' is not that of some infinite and infinitely divisible space, the space of a receptacle, that the intellect posits as the place in which and from which everything is built.

It would be mistaken to think that Bergson's writings simply reinstate, in some tired old and predictable fashion, the superiority of philosophy over science or that he condemns science to lesser work on account of holding on to some *a priori* conception of the limits of science.[41] In *Matter and Memory* he speaks of a new alliance between science and philosophy taking place around a thinking beyond the human condition, in which would be discovered 'the natural articulations of a universe we have carved artificially.'[42] In his 'Introduction to Metaphysics' he actually goes so far as to suggest that science and metaphysics 'meet in intuition', and he looks forward to a future praxis of thinking in which there is more science put into metaphysics and more of metaphysics put into science.[43] Of course, by metaphysics here Bergson means something quite precise and specific, as well as something novel and creative, namely, a mode of thinking that goes beyond the human condition. The science to be critiqued for Bergson is that mode of knowing which carries with it an *'unconscious* metaphysic', by which he means a metaphysics to which it is oblivious, and which it is unable to work through (Dennett's Darwinism provides us with an excellent example of this). For Bergson the history of science is not to be thought in terms of some simple linear narrative of making greater and greater progress towards the 'truth'. This is not because he is unconcerned with truth – he undoubtedly is; rather, his concern is with the 'absolute' (with knowledge of life that goes beyond the 'relativity' of our human habits of understanding), and the history of science demands a more complicated reading from us in terms of detecting those 'soundings in pure duration' that punctuate it. The achievement in science of such 'soundings' takes us beyond the restrictive empiricism Bergson often places, illegitimately, upon its activity.

Bergson defines the task of philosophy as one of thinking beyond the human condition. While such a task may seem deeply paradoxical, there is nothing mysterious about it. By the phrase 'the human condition' is meant something quite specific, namely, the mental and social habits that have informed human evolution and prevent us, owing to their essentially spatialised character and utilitarian impulse, from thinking a 'creative evolution'. For Bergson the condition of the intellect is, in fact, one that conforms to certain tendencies of matter itself, which is why he is keen to demonstrate a *double genesis* of both matter and the intellect. The intellect conditions our mental habits and informs our social practices to the detriment of *life* (as Bergson points out, it 'dislikes what is fluid and solidifies everything it touches'). But it is life that continues to live through us and that transcends the intellect. This essential Bergsonian insight informs Deleuze's commitment to a philosophical practice in which, 'Life will no longer be made to appear before the categories of thought; thought will be thrown into the categories of life.'[44] This inversion continues to pose a challenge to a philosophical thinking of life. The questions is: does it pose a challenge for science as well?

Department of Philosophy
Faculty of Social Sciences
University of Warwick
Coventry, CV4 7AL, UK

Notes

1 In Henri Bergson, *Creative Mind*, trans. M.L. Andison (New York: Citadel Press, 1992), p. 193 [Bergson, *Œuvres*, ed. A. Robinet (Paris: Presses Universitaires de France, 1959), p. 1425].

2 Henri Bergson, *Creative Evolution*, trans. A. Miller (Lanham: University Press of America, 1983), p. 340 [*Œuvres*, p. 782]. Cited hereafter as CE/Œ.

3 Alain Badiou, 'Gilles Deleuze, *The Fold: Leibniz and the Baroque*', in C.V. Boundas & D. Olkowski (eds.), *Gilles Deleuze and the Theatre of Cruelty* (New York: Routledge, 1994), pp. 51–73, p. 65.

4 Gaston Bachelard, *The Dialectic of Duration*, trans. M. McAllester Jones (Manchester: Clinamen Press, 2000), p. 24 [*La dialectique de la durée* (Paris: Presses Universitaires de France, 1936), p. 2].

5 Gilles Deleuze, *Bergsonism*, trans. H. Tomlinson & B. Habberjam (New York: Zone Books, 1991), p. 34 [*Le bergsonisme* (Paris: Presses Universitaires de France, 1966), p. 27]).

6   See Bergson, *Two Sources of Morality and Religion*, trans. R. Ashley Audra & C. Brereton (Notre Dame: University of Notre Dame Press, 1977), p. 112 (*Œuvres*, p. 1069).
7   *Bergsonism*, p. 38 (*Le bergsonisme*, p. 31); for further discussion of Bergson's appropriation of the Riemannian notion of the multiplicity to think duration, see R. Durie, 'Splitting Time: Bergson's Philosophical Legacy', *Philosophy Today*, 44 (2000), pp. 152–168.
8   Bergson, *Time and Free Will*, trans. F.L. Pogson (London & New York: Macmillan, 1910), pp. 121–3 [*Œuvres*, pp. 80–2]; Deleuze, *Bergsonism*, pp. 38–43 [*Le bergsonisme*, pp. 31–7].
9   Deleuze, *Bergsonism*, p. 42 [*Le bergsonisme*, p. 36].
10 Life is spoken of in terms of an impetus simply because 'no image borrowed from the physical world can give more nearly the idea of it.' It provides us with insight into the evolution of life as the enfolding of a plurality of interpenetrating terms. If we view life in terms of its contact with matter then it is an impetus. But regarded in itself, it is 'an immensity of virtuality', that is, a 'mutual encroachment of thousands and thousands of tendencies' (and, let us note, they are only 'thousands and thousands' when we think them spatially!). [CE 258/Œ 714]
11 See H. Bergson, *Duration and Simultaneity* (Manchester: Clinamen Press, 1999) p. 33. For further insight see the excellent Introduction to the new edition by Robin Durie, pp. vii–xxvi. For my own treatment of the relation between Bergsonian time and the times of Relativity se the opening essay in my *Philosophy and Virtual Life: Time, Consciousness, Evolution, and Event* (London: Routledge, forthcoming).
12 Bergson, *Creative Mind*, p. 18 [*Œuvres*, p. 1260]; translation modified.
13 On the glass of sugared water, see *Creative Evolution*, pp. 10 & 339 [*Œuvres*, pp. 502 & 781–2]; *Creative Mind*, pp. 20–1 [*Œuvres*, p. 1262]; and Gilles Deleuze, *Cinema 1: The Movement-Image*, trans. H. Tomlinson & B. Habberjam (London: Athlone, 1983) pp. 10–11 (*Cinema 1. L'Image-mouvement* [Paris: Les Editions de Minuit, 1983], pp. 20–1).
14 *Creative Mind*, p. 21 [*Œuvres*, p. 1262].
15 Bergson, 'The Possible and the Real', in *Creative Mind*, p. 105 [*Œuvres*, p. 1344].
16 Compare M. Eigen, *Steps Towards Life. A Perspective on Evolution* (Oxford: Oxford University Press 1992), p. 3: 'Life is *not* an inherent property of matter'.
17 Eigen, *Steps Towards Life*, p. 123.
18 'The Possible and the Real', p. 95 [*Œuvres*, p. 1335].
19 *Creative Mind* p. 100 [*Œuvres*, p. 1339].
20 As Deleuze points out, duration is 'real succession, but it is so only because, more profoundly, it is *virtual coexistence*.' [*Bergsonism*, p. 60/*Le bergsonisme*, p. 56]. Although Deleuze's reading of Bergson is rightly celebrated for

drawing out the significance of this thinking of the co-existence of past and present, it can also be found articulated in similar terms in Karin Stephen's excellent early study, *The Misuse of Mind. A Study of Bergson's Attack on Intellectualism* (London: Kegan Paul, 1922), pp. 70ff. Stephen's point is that past and present cannot be conjoined in terms of 'external relations'.

21 Bergson identifies such a Platonism as informing the ambitions of Kant's *Critique of Pure Reason* in its quest for a universal mathematics. It is worth citing him at length on this point: 'Universal mathematics is what the world of Ideas becomes when one assumes that the Idea consists in a relation or a law, and no longer in a thing... The main task of the *Critique*, therefore, was to lay the foundations of this mathematics, to determine what the intelligence should be and what should be the object in order that an unbroken mathematics might bind them together. And it follows that if all possible experience is thus assured of admittance into the rigid and pre-constituted frameworks of our understanding (unless we assume a pre-established harmony), our understanding itself organizes nature and finds itself reflected in it as in a mirror. Whence the possibility of science, which owes all its effectiveness to its relativity – and the impossibility of metaphysics, which finds it has nothing more to do than to parody, on the phantom of things, the work of conceptual arrangement which science pursues seriously on relations. In short, *the whole* Critique of Pure Reason *leads to establishing the fact that Platonism, illegitimate if Ideas are things, becomes legitimate if ideas are relations, and the ready-made idea, once brought down from heaven to earth, is indeed as Plato wished, the common basis of thought and nature. But the whole* Critique of Pure Reason *rests also upon the postulate that our thought is incapable of anything but Platonizing, that is, of pouring the whole of possible experience into pre-existing molds.'* [*Creative Mind*, pp. 196–7 (*Œuvres*, pp. 1428–9); translation modified]

22 G. Adamson, 'Henri Bergson: Evolution, Time and Philosophy', in *World Futures* 54 (1999), p. 136.

23 For further insight into Bergson and finality see Keith Ansell Pearson, 'Finality and Virtuality: Bergson's Response to Kant', in *Tekhnema: A Journal of Philosophy and Technology* (forthcoming, 2000).

24 In his essay on 'Laughter' Bergson actually ascribes a subversive and interruptive role to mechanism within life with the aid of the notion of 'absent-mindedness' [Henri Bergson, *Laughter: An Essay on the Meaning of the Comic*, trans. C. Brereton & F. Rothwell (Kobenhavn & Los Angeles: Green Integer, 1900), p. 139 (*Œuvres*, p. 461)].

25 Daniel C. Dennett, *Darwin's Dangerous Idea: evolution and the meanings of life* (London: Allen Lane, 1995), p. 49. [Cited hereafter as DDI].

26 Transcendental not in the sense of a condition of *our* mode of knowing something but in the sense of rendering a notion of evolution or becoming intelligible.

27 Adamson, 'Henri Bergson: Evolution, Time and Philosophy', p. 144.

28 Bergson, *Creative Mind*, p. 26 [*Œuvres*, p. 1267–8].
29 See Badiou, 'Gilles Deleuze, *The Fold: Leibniz and the Baroque*', and *Deleuze: The Clamor of Being*, trans. L. Burchill (Minneapolis: University of Minnesota Press, 2000). I have examined Badiou's thinking of the event, in relation to both Bergson and Deleuze, in Ansell Pearson, 'Thinking Immanence: On the Event of Deleuze's Bergsonism', in G. Genosko (ed.), *Deleuze and Guattari: Critical Assessments* (London: Routledge, 2000).
30 Bachelard, *The Dialectic of Duration*, p. 46 [*La dialectique de la durée*, p. 28]. For an account of the paradoxes of set theory, and insight into their relation to the paradoxes of Zeno, see the opening chapter of M. Tiles, *The Philosophy of Set Theory: An Historical Introduction to Cantor's Paradise* (Oxford: Basil Blackwell, 1989), pp 10ff; see also the study by A.W. Moore, *The Infinite* (London: Routledge, 1990).
31 For a discussion of Leibniz's use of the notion of the fold to reconcile the apparent contradiction between the law of continuity and the contiguity inherent in the principle of indiscernables, see Gilles Deleuze, *The Fold: Leibniz and the Baroque*, trans. T. Conley (London: Athlone, 1993), pp 20ff.
32 *Bergsonism*, p. 133 [*Le bergsonisme*, p. 104].
33 Deleuze, *Cinema 1*, p. 9 (*Cinema 1. L'Image-mouvement*, p. 20).
34 Bachelard, *The Dialectic of Duration*, p. 29 [*La dialectique de la durée*, p. 8].
35 Cf. 'The Possible and the Real', p. 93 [*Œuvres*, p. 1333].
36 Bachelard, *The Dialectic of Duration*, p. 29 [*La dialectique de la durée*, p. 8].
37 Bachelard, 'The Instant', p. 71 of the present volume [*L'Intuition de l'instant* (Paris: Editions Stock, 1931), p. 23]
38 On the role of contingency within Bergson's account of creative evolution, see *Creative Evolution*, pp. 255ff [*Œuvres*, pp. 711ff]: 'The part played by contingency in evolution is therefore great. Contingent are the forms adopted, or rather invented. Contingent, relative to the obstacles encountered in a given place and at a given moment, is the dissociation of the primordial tendency into such and such complementary tendencies which create divergent lines of evolution. Contingent the arrests and set-backs; contingent, in large measure, the adaptations.'
39 P. Grasse, *Evolution of Living Organisms* (New York: Academic Press, 1977), p. 19.
40 Bergson, *Matter and Memory*, p. 197 [*Œuvres*, p. 333].
41 Prigogine and Stengers suggest that his error lay in attributing 'to science *de jure* limitations that were only *de facto*'. [Ilya Prigogine & Isabelle Stengers, *Order out of Chaos* (London: Fontana, 1985), p. 93].
42 *Matter and Memory*, p. 197 [*Œuvres*, p. 333].
43 See *Creative Mind*, pp. 192–200 [*Œuvres*, pp. 1424–32].
44 Gilles Deleuze, *Cinema 2: The Time-Image*, trans. R. Galeta & H. Tomlinson (London: Athlone Press, 1985), p. 189 [*Cinema 2. L'Image-temps* (Paris: Les Editions de Minuit, 1985), p. 246].

8
The Plane of the Present and the New Transactional Paradigm of Time

John G. Cramer

1. Introduction.
Time and the Plane of the Present
as Evolving Paradigms

The *plane of the present* is a concept that is useful for discussing the various paradigms of time. Here by 'plane of the present' we mean the temporal interface that represents the present instant and that forms the boundary between the past and the future. We use the geometrical term 'plane' to indicate an extended surface in the space-time continuum, as opposed to a 'point' on some time axis. This point/plane dichotomy is intended to raise issues of extension and simultaneity and to examine the degree to which these are meaningful concepts from various physical viewpoints. We will show by example in the present work that the plane of the present is a pivotal concept that offers considerable power in differentiating between various views of the nature of time.

The concept of time within the main stream of physics thinking has followed a rather convoluted path over the past three millennia. Anticipating the modern motion picture, Zeno of Elea (c.490–c.430 B.C) questioned whether time should appropriately be viewed as a continuously flowing river, or should more properly be considered as a rapid sequence of stop-motion 'freeze-frames', in effect rendering geometrical each instant as a separate infinitesimal point on the line of time. Adopting this view, he asked how physical motion could occur. He argued paradoxically that motion is not possible, since it appears to happen only *between* the frozen frames of time instants.[1]

From the viewpoint of Zeno, the plane of the present would be simply the last and most recent in this sequence of freeze-frames. It would be that frozen instant, spanning the universe, which changes progressively as the instant we call 'now' becomes the frozen past and future possibility freezes into the 'now' of present reality. We note that the plane of the present as a concept does not resolve the arrow paradox that Zeno raised. It only provides a way of thinking about it.

Isaac Newton (1642–1727) changed the paradigm of time by introducing the concept of absolute time, a sequence of instants that are the same at all points in space. Thus the present is a locus of a universal time characterizing a three dimensional volume of space at some instant during its dynamic evolution. For Newton, time was an absolute clock that ticked everywhere at the same rate. He altered and extended Zeno's instant of time by inventing differential calculus, thereby transforming the frozen time frames to *time derivatives*, the essential embodiment of change in the infinitesimal limiting case. In the process, Newton implicitly provided the solution to Zeno's arrow paradox by providing a formalism in which motion and change are explicitly a part of each freeze frame.

The mathematical concept of a series expansion around an instant of time, a later extension of Newton's ideas, implies that the entire past and future of any continuous functional behavior is embodied in the values of the time-dependent function and all of its derivatives, evaluated only at one single instant of time. This led to the 19th century paradigm of a 'clockwork universe' in which, given the positions and velocities of all particles at some instant of time and a knowledge of their interaction forces, the entire past and future of the universe was determined and, in principle, could be calculated.

In such a deterministic universe, the plane of the present loses much of its significance. The present is simply the location of the 'bead' representing our consciousness as it slides along a fixed wire that spans the frozen past and the pre-determined future.

Albert Einstein (1879–1955) changed the time paradigm yet again, rejecting Newton's absolute universal time and instead depicting time as mutable, a fourth dimension that in some sense is interchangeable with the three spatial dimensions. Each inertial reference frame (i.e., observer moving with some constant velocity), in Einstein's view, achieves its own particular trade-off between the local space and time dimensions. In this democracy of dimensions, time is placed on a more equal footing with space. This led to the concept of a 'block universe', an infinite stack of three dimensional infinitesimal time-slices of the universe, in which all regions of time, future as well as past, must co-exist in the overall four-space continuum. Each observer-based reference frame makes its own angled time-slice through this block, representing the view of that observer as to which event-points in the four-dimensional block occur at the same time.

With special relativity Einstein destroyed the Newtonian concept of absolute time. In Einstein's world it is not meaningful to speak of two spatially separated events as truly 'simultaneous'. Instead, the relative time of occurrence of separated space-time events depends on the reference frame of the observer. The reference frame also affects the rate at which time progresses, so that time is slowed in a moving reference frame, with the question of which frame is moving also depending on the observer.

The plane of the present in the Einsteinian view is thus not an immutable surface that always cuts through space-time in the same way, but rather a mutable plane that may tilt at an angle that depends on the observer's reference frame. Moreover, Einstein showed, with the introduction of general relativity, that time slows, and indeed may stop altogether, in a sufficiently strong gravitational field. Therefore, time may run at different rates in locations with different strengths of the local gravitational field. The concept of 'proper time' in a reference frame is all that remains of Newtonian time. Some plane of the present (or surface of simultaneity) may connect events in different regions of this time-distorted space-time, but it has become an elusive and convoluted concept.

Werner Heisenberg (1901–1976) again changed the time paradigm with his formulation of quantum mechanics, the physics theory of matter and energy at the smallest scales.[2] Heisenberg found that the *uncertainty principle* was contained within his new quantum formalism.[3] The uncertainty principle is in essence a new physical law showing that there is a specific minimum uncertainty between particular pairs of 'complementary' physical variables. These uncertain pairs of complementary variables include position and momentum (along any coordinate axis) and (of particular relevance for the present discussion) time and energy.

The uncertainty principle demonstrates that the concept of the instant is a geometrical ideal rather than a physical concept, because the uncertainties in the determinations of time and energy are inextricably entangled within the time-energy uncertainty relation. Any measurement of energy and/or any constraints on energy in a physical process will unavoidably bring with it an uncertainty in the time variable. Consequently, since some information about energy is always available in any real physical process, the concept of an infinitesimal instant of time is a geometrical idealization that cannot be applied to the physical universe. Similarly, a precise measurement of velocity (or momentum, to be precise) brings with it an unavoidable uncertainty in position, so the concept of a mathematical point representing position is also inappropriate.

The Newtonian clockwork universe is thus destroyed by the uncertainty principle. Heisenberg's indeterminacy means that the past and future are not determinable from the data of an instant. The future, outside a very limited region, is uncertain and unpredictable.

Correspondingly, the plane of the present, instead of being the perfectly sharp edge implied by the Newtonian view, is a boundary that is blurred by the uncertainty principle. In the quantum world, the plane of the present must be a fuzzy region with the past on one side in time and the future on the other, but with some uncertainty in the central region about which is the past and which is the future. Across this region there must occur the freezing of possibility into reality, but possibility and actuality are mixed and smeared by time indeterminacy.

As we have seen from the preceding discussion, the evolution of the paradigm of time was driven not by the speculations of philosophers but by physics, the activity of checking concepts against the structure of the universe. This process has periodically required revision of our ideas, as our ability to do such checking has improved. This is an ongoing process, the next step of which may come from an interpretation of quantum mechanics.

2. The Transactional Interpretation of Quantum Mechanics

Quantum mechanics is unique among the major theories of physics in having emerged from the mathematical formulations of Schrödinger and Heisenberg in the 1920s without being accompanied by a clear picture of the physical processes described by the mathematical formalism. The mathematical formulation of quantum mechanics is now well established and tested, but there is still great controversy concerning the *interpretation* of that formalism.[4]

In the formalism of quantum wave mechanics, a physical system is represented by a second-order differential equation or 'wave equation'. Maxwell's electromagnetic wave equation and the Schrödinger equation are two examples of wave equations used in quantum mechanics. Within the wave equation, the physical properties of a system are represented by boundary conditions that characterize known limits and regions of applicability and by a potential that characterizes the way in which energy changes when elements of the system are moved or displaced. The wave equation is solved mathematically to produce a solution called

The Plane of the Present

a 'wave function'. The wave function is then used, applying standard quantum mathematical procedures, to make predictions about the probable outcomes (or 'expectation values') of physical measurements that might be made on the system, e.g., its position, momentum, energy, etc.

In the 1920s when quantum mechanics was introduced, there was perceived to be a serious problem. The two major physical theories developed in the first decades of the 20th century, special relativity and quantum mechanics, were initially viewed as incompatible. This incompatibility arose because the Schrödinger equation, the wave equation used to describe the behavior of particles of non-zero rest mass, is inconsistent with relativity. Fortunately, later developments in quantum mechanics produced alternative wave equations, the Klein-Gordon equation for integer-spin bosons and the Dirac equation for half-integer spin fermions, which are fully consistent with relativity. Further, it was found that the Schrödinger equation could be viewed as the non-relativistic (low-velocity) limit of either of these more correct wave equations.[5]

While the wave function has proved to be an extremely useful mathematical object for applications of quantum mechanics, its *meaning* remains a matter of heated debate. In classical mechanics, such a wave function would represent a wave physically present in space (for example, a traveling electromagnetic radio wave), but in quantum mechanics this simple explanation of the wave function is conventionally rejected. Since the late 1920s the orthodox view, as embodied in the Copenhagen interpretation of quantum mechanics, has been that a wave function is *not* physically present in space, but rather is a purely mathematical construct used for calculation, which can be interpreted as a mathematical function encoding the state of knowledge of some observer. While the Copenhagen view is self-consistent, it prevents the visualization of quantum mechanical processes. Further, careful considerations of the implications of the Copenhagen interpretation have generated a number of 'interpretational paradoxes',[6] e.g., Schrödinger's Cat, Wigner's Friend, Wheeler's Delayed Choice, the Einstein-Podolsky-Rosen paradox, and so on, that lead into very deep philosophical waters.

The *transactional interpretation of quantum mechanics*,[7] originally presented in a long review article in 1986, is an alternative to the Copenhagen interpretation that avoids these paradoxes. It is relativistically invariant, so that it can be used with the relativistic wave equations discussed above, and it uses the advanced and retarded wave function solutions of these equations in a 'handshake' that provides a rationale for treating the wave function as physically present in space.

The logical development of the transactional interpretation starts with the time-symmetric classical electromagnetism of Dirac,[8] and Wheeler and Feynman[9] which describes electromagnetic processes as an exchange between retarded (normal) and advanced (time-reversed) electromagnetic waves. The transactional interpretation applies the time-symmetric Wheeler-Feynman view to the quantum mechanical wave function solutions of the electromagnetic wave equation. The lessons learned about electromagnetic quantum waves are then extended to wave functions describing the behavior of massive particles (e.g., electrons and protons) by applying the same interpretation to their relativistic wave equations. Finally, the Schrödinger equation is included as a non-relativistic reduction of the relativistic wave equations in the limit of small velocities.

The transactional interpretation views each quantum event as a 'handshake' or 'transaction' process extending across space-time that involves the exchange of advanced and retarded waves to enforce the conservation of certain quantities (energy, momentum, angular momentum, etc.). It asserts that each quantum transition forms in four stages: (1) *emission*, (2) *response*, (3) *stochastic choice*, and (4) *repetition to completion*.

The first stage of a quantum event is the *emission* of an 'offer wave' by the 'source', which is the object supplying the quantities transferred. The offer wave is the time-dependent retarded quantum wave function Ψ, as used in standard quantum mechanics. It spreads through space-time until it encounters the 'absorber', the object receiving the conserved quantities.

The second stage of a quantum event is the *response* to the offer wave by any potential absorber (there may be many in a given event). Such an absorber produces an advanced 'confirmation wave' Ψ^*, the complex conjugate of the quantum offer wave function Ψ. The confirmation wave travels in the reverse time direction and arrives back to the source at precisely the instant of emission with an amplitude of $\Psi\Psi^*$.

The third stage of a quantum event is the *stochastic choice* exercised by the source in selecting one from among the possible transactions. It does this in a linear probabilistic way based on the strengths $\Psi\Psi^*$ of the advanced-wave 'echoes' it receives from the potential absorbers.

The final stage of a quantum event is the *repetition to completion* of this process by the source and absorber, reinforcing the selected transaction repeatedly until the conserved quantities are transferred and the potential quantum event becomes real.

Since the advanced-retarded-wave handshake used by the transactional interpretation operates in both time directions, it is in a sense atemporal, in that no elapsed time at the space-time site of the emitter is required between the beginning of a quantum event as an offer wave and its conclusion as a completed transaction (or 'collapsed wave function', in the terminology of the Copenhagen interpretation). Similarly, at the space-time site of the absorber (the future end of a transaction), there is no elapsed time between responding with a confirmation wave and the completion of the transaction. The transactional interpretation asserts that at the quantum level time is a two-way street, in which at some level the future determines the past as well as the past determining the future.

However, there is time direction preference associated with quantum mechanics. In the formalism of quantum mechanics, processes are usually analyzed using what is called the 'post' formalism. In terms of the transactional interpretation, this is the assumption that the probability of a quantum transition $\Psi\Psi^*$ is evaluated at the space-time location of the emitter (the 'past' end of the transaction) rather than at the location of the absorber (the 'future' end of the transaction). The alternative is the time-reversed 'prior' formalism, in which the probability is evaluated at the future absorber end of the process.[10] Normally in quantum calculations the two formalisms, if calculated with sufficient accuracy, give the same result, but there are cases involving violations of time reversal invariance where they do not. Therefore, the transactional interpretation, despite its even-handed treatment of advanced and retarded waves, implies a limited preference for a time direction on the basis of its description of transaction formation.

3. The Hierarchy of the Arrows of Time

In the situation described above in which quantum wave functions travelling in both time directions are handled in an even-handed and symmetric way, one must ask about the origins of the macroscopic 'arrow of time' that is evident in the everyday world. This question is complicated by the presence of at least five seemingly independent 'arrows of time' that can be identified in the physical world. Let me briefly review them.

1. **Subjective:** At the macroscopic level, it is self-evident that the past and the future are not the same. We remember the past but not the future. Our actions and decisions can affect the future but not the past.
2. **Electromagnetic:** We can send electromagnetic signals to the future but not to the past. A current through an antenna makes retarded positive-energy waves but not advanced negative-energy waves.

3. **Thermodynamic:** Isolated systems have low entropy (i.e., disorder) in the past and gain entropy and become more disordered in the future. Molecules released from a confining box will rapidly fill a larger volume, but they will not spontaneously collect themselves back in the box.
4. **Cosmological:** The universe was smaller and hotter in the past but will be larger and cooler in the future. The time direction of expansion is an arrow.
5. **CP Violation:** The $K^°_L$ meson (the neutral long-lived K meson, which is a matter-antimatter combination of a down quark and a strange quark) exhibits weak decay modes having matrix elements and transition probabilities that are larger for the decay process than for the equivalent time-reversed process, in violation of the principle of time-reversal invariance. This is related to the so-called CP violation[11] that is present in the neutral K meson system, which shows a preference for matter over antimatter in certain decay processes.

These indications of time asymmetry cannot be independent, and one would like to understand their connections and hierarchy. It is generally agreed that the CP violation arrow is probably the most fundamental time arrow and is responsible for both the dominance of matter in the universe and the breaking of time symmetry in fundamental interactions to make possible the expansion of the universe in a particular time direction, thereby leading to the cosmological arrow of time. From this point on, however, there are at least two divergent views of the hierarchy.

The first of these views might be characterized as the orthodox view of hierarchy, in that it has been widely advocated in the physics literature by many authors, particularly Hawking.[12] It is illustrated by Figure 1(a) and asserts that the thermodynamic arrow follows from the cosmological arrow and leads to the electromagnetic and subjective arrows of time.

This view goes back to Ludwig Boltzmann (1844–1906), who in the 1870s 'derived' from first principles the thermodynamic arrow of time and the 2nd law of thermodynamics with his famous H-Theorem. If, as Boltzmann appears to demonstrate, the increase in entropy can be derived from first principles using only statistical arguments, this would provide strong support for its primary position in the hierarchy of time arrows.

However, the 'first principles' that lead to Boltzmann's H-Theorem include an implicit time-arrow. Boltzmann used the apparently reasonable assumption that the motions of colliding members of a system of

The Plane of the Present

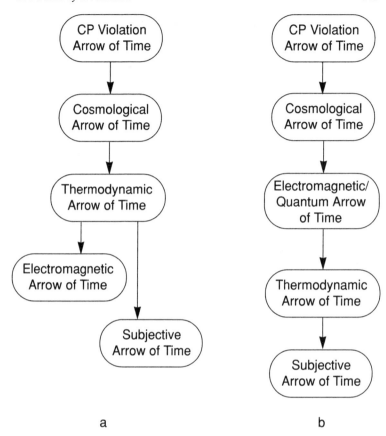

Figure 1. Two alternative hierarchies of the arrows of time: (a) the hierarchy generally favored in the physics literature; and (b) the hierarchy advocated here.

particles are uncorrelated *before a collision*. Unfortunately, that assumption is not as innocent as it appears. It smuggles into the problem an implicit time asymmetry, which ultimately leads to a system entropy that is constant or increasing with time. However, if one, in the spirit of extending the H-theorem, had assumed that the motions of colliding particles were uncorrelated *after the collision*, then one would have demonstrated with equal rigor that the entropy was *constant or decreasing* with time, i.e., the thermodynamic arrow would be pointing in the wrong direction.

This consideration leads to the view of the time arrow hierarchy illustrated by Figure 1(b), which asserts that the electromagnetic arrow of time produces the thermodynamic arrow because of the dominance of retarded electromagnetic waves and interactions in the universe. The source of the time asymmetry implicit in the assumptions of the H-Theorem is the intrinsic 'retarded' character of electromagnetic interactions. Here the term retarded means that there is a speed-of-light time delay between the occurrence of some change in a source of electromagnetic field, i.e., the movement of an electric charge, and its appearance in distant electromagnetic fields produced by that source. The field change always occurs *after* the source change, in any reference frame.[13]

The time arrows listed above did not include a quantum mechanical arrow of time. However, we have observed that from the transactional viewpoint the present source or emitter selects from among the possible quantum transactions, not the future absorber. How does this time preference fit into the above hierarchy?

In effect, this quantum arrow of time is equivalent to the electromagnetic arrow, which requires the dominance of retarded waves. The conditions of our universe favor retarded waves and positive energies. While quantum processes may include the actions of advanced waves to enforce conservation laws, no net 'advanced effects' are allowed. Such effects would represent violations of the principle of causality, and have not been observed. We conclude that the hierarchy indicated in Figure 1(b), the primacy of the electromagnetic/quantum arrow, is consistent with the quantum time preference discussed above and can be used along with a more enlightened version of Boltzmann's H-Theorem to derive the thermodynamic arrow of time.

4. Determinism and the Transactional Interpretation

It has been asserted[14] that the transactional interpretation is necessarily deterministic, requiring an Einsteinian block universe to pre-exist, because the future must be fixed in order to exert its influence on the past in a transactional handshake. However, while block-universe determinism is consistent with the transactional interpretation, it is not required. A part of the future is emerging into a fixed local existence with each transaction, but the future is not determining the past, and the two are not locked together in a rigid embrace.

Let us make an analogy. The handshakes envisioned by the transactional interpretation bear some resemblance to the handshakes that

The Plane of the Present 187

take place on telephone or Internet lines these days when one uses a debit card to make a purchase at a shop. The shop's computer system reads the magnetic strip on your card and transmits the information to the bank, which verifies that your card is valid and that your bank balance is sufficient for the purchase, and then removes the purchase amount from your bank balance. This transaction enforces 'conservation of money'. There is a one to one correspondence between the amount the store receives for your purchase and the amount that is deducted from your bank account. The transaction assures that the amount is deducted only once, from only one bank account and credited only once, to only one store. On the other hand, the bank does not exert any influence on what you choose to purchase, beyond ensuring that money is conserved in the transaction and that you do not overspend your resources.

A quantum event as described by the transactional interpretation follows the same kind of protocol. There is a one to one correspondence between the energy and other conserved quantities (momentum, angular momentum, spin projections, etc.) that are conveyed from the emitter to the absorber, but aside from the enforcement of conservation laws, the future absorber does not influence the past emission event. Therefore, the 'determinism' implied by the transactional interpretation is very limited in its nature.[15]

5. The Plane of the Present and the Transactional Interpretation of Quantum Mechanics

This brings us to the question of how our conceptualization of the plane of the present is affected by the transactional interpretation. In the transactional interpretation, the freezing of possibility into reality as the future becomes the present is not a plane at all, but a fractal-like[16] surface that stitches back and forth between past and present, between present and future.

To make another analogy, the emergence of the unique present from the future of multiple possibilities, in the view of the transactional interpretation, is rather like the progressive formation of frost crystals on a cold windowpane. As the frost pattern expands, there is no clear freeze-line, but rather a moving boundary, with fingers of frost reaching out well beyond the general trend, until ultimately the whole window pane is frozen into a fixed pattern. In the same way, the emergence of the present involves a lacework of connections with the future and the past, ensuring that the conservation laws are respected and the balances of energy and momentum are preserved.

Is free will possible in such a system? I believe that it is. Freedom of choice does not include the freedom to choose to violate physical laws. The transactional handshakes between present and future are acting to enforce physical laws, and they restrict the choices between future possibilities only to that extent.

Therefore, we conclude that the transactional interpretation does not require a deterministic block universe. It does, however, imply that the emergence of present reality from future possibility is a far more complex process than we have previously been able to imagine. This is the new transactional paradigm of time.

Department of Physics
University of Washington
Seattle, Washington, USA

Notes

1 We note that modern quantum optics has provided a new demonstration of Zeno's viewpoint, with the freeze-frames taking the form of the collapse of the quantum wave-function following measurement and the continuous time evolution occurring with the absence of measurements (see P.G. Kwiat, et al., 'High-efficiency Quantum Interrogation Measurements via the Quantum Zeno Effect', *Phys. Rev. Letters* 83, (1999) 4725–4728; also available on the Internet at http://xxx.lanl.gov/quant-ph/9909083). Because a quantum measurement of a time-evolving system can 'reset' the system to its initial state repeatedly, thereby blocking the time evolution that would otherwise occur, the repeated measurements can freeze Zeno's arrow in one particular freeze-frame and prevent its motion.

2 W. Heisenberg, 'On the Quantum Interpretation of Kinematical and Mechanical Relationships', *Zeitschrift für Physik* 33 (1925), 879–893.

3 W. Heisenberg, 'The Actual Content of Quantum Theoretical Kinematics and Mechanics', *Zeitschrift für Physik* 43 (1927), 172–198.

4 There are actually two quantum formalisms, the matrix mechanics of Heisenberg and the wave mechanics formalism of Schrödinger and Dirac. However, since these have been shown to be completely equivalent, we will focus on Schrödinger-Dirac wave mechanics.

5 We note that quantum field theory, the most widely used formalism in high energy physics, is consistent with special relativity and avoids wave equations altogether, while the task of making quantum mechanics compatible with *general relativity*, the current standard model of gravity, remains an

unsolved problem.
6 J.G. Cramer, 'The Transactional Interpretation of Quantum Mechanics', *Reviews of Modern Physics* 58 (1986), 647–687; also available on the Internet at http://faculty.washington.edu/jcramer/theory.html.
7 J.G. Cramer, 'The Transactional Interpretation of Quantum Mechanics'; J.G. Cramer, 'An Overview of the Transactional Interpretation of Quantum Mechanics', *International Journal of Theoretical Physics* 27 (1988), 227–236; also available on the Internet at http://faculty.washington.edu/jcramer/theory.html.
8 P.A.M. Dirac, 'Classical Theory of Radiating Electrons', *Proc. Royal Soc. London* 167A (1938), 148–169.
9 J.A. Wheeler & R.P. Feynman, 'Interaction with the Absorber as the Mechanism of Radiation', *Reviews of Modern Physics* 17 (1945), 157–181; J.A. Wheeler & R.P. Feynman, 'Classical Electrodynamics in Terms of Direct Interparticle Action', *Reviews of Modern Physics* 21 (1949), 425–434.
10 R.M. DeVries, G.R. Satchler, and J.G. Cramer, 'Importance of Coulomb Interaction Potentials in Heavy-Ion Distorted Wave Born Approximation Calculations', *Phys. Rev. Letters* 32 (1974), 1377.
11 In quantum mechanics, CP is a symmetry transformation involving the simultaneous conversion of matter particles to antimatter and vice versa (C) and the reversal of the three spatial coordinate axes (P).
12 Stephen W. Hawking, 'Arrow of Time in Cosmology', *Phys. Rev.* D32 (1985), 2489.
13 For further discussion of the arrow of time problem, see J.G. Cramer, 'The Arrow of Electromagnetic Time and Generalized Absorber Theory', *Foundations of Physics* 13 (1983), 887–902; also available on the Internet at http://faculty.washington.edu/jcramer/theory.html; and J.G. Cramer, 'Velocity Reversal and the Arrows of Time', *Foundations of Physics* 18 (1988), 1205–1212; also available on the Internet at http://faculty.washington.edu//jcramer/theory.html.
14 J.F. Woodward, 'Making the Universe Safe for Historians; Time Travel and the Laws of Physics', *Foundations of Physics Letters* 8 (1995), 1–40.
15 The same kind of limited connections and information transfer, for similar reasons, is present in so-called EPR experiments involving measurements on entangled quantum subsystems. The non-local quantum connection between a pair of separated measurements preserves correlations and enforces conservation laws without permitting the transfer of information from observer to observer.
16 The time-interface surface cannot be truly fractal in its structure because the scale terminates at the quantum limit. A quantum system cannot be 'turtles all the way down'.

9
The Complexity of the Instant:
Bachelard, Levinas, Lucretius[1]

David Webb

> *Le temps ne dure qu'en inventant.*
> Gaston Bachelard[2]
>
> An instant and a minimum deviation are enough.
> Michel Serres[3]

If time is discontinuous, how does each instant give way to the next? When does each instant end and the next begin? The idea that time might be composed out of instants, each placed one alongside the next, was exposed as incoherent by Aristotle. Since then, time has been treated as fundamentally continuous and the instant as an abstraction. On this view the continuity of time reflects the continuity of change and of the ontological order articulated in change. The introduction of non-linearity and complexity into the conception of change in modern science has demanded a revision of these assumptions, presenting a fresh and compelling challenge to the way we think about time. We must discover in time not just the unfolding of events, but the occurrence of what is *radically new*, in the sense that it was not prefigured in advance by a potentiality or by the application of physical law to the initial conditions of a system.

The relation of time to change has meant that our understanding of time has been developed in tandem with our conception of sensible experience and has in general been shaped by that conception. In different ways, Bachelard and Levinas both press this association of time with *sensibility* to its limits. They both reverse the traditional bias towards continuity by beginning from the instant, and both treat the instant as the articulation of a discontinuity that allows for relations of *complexity* and *chance* that are suppressed by our usual conception of time. Discontinuity in this sense is less the absence of relation than the original event of a new relation, and in particular, the relation between consciousness and reality. In all of this, Bachelard is indebted to Ancient Greek atomism, which provides him with an immediate counterpoint to the classical tradition founded on the physics and metaphysics of Aristotle and Plato. While Levinas' work is indebted above all to the phenomenology of Husserl, he also borrows from Bachelard, and especially those elements of Bachelard's account of time and the instant that derive from atomism. One of the central tenets of atomism was its admittance of the void alongside being and each of the readings of the instant we shall review differs over how the void is to be treated, and the implications of this treatment for thinking about time. Despite these differences, I shall argue that both Bachelard and Levinas are bound by constraints that prevent their accounts of the complexity of the instant from contributing to an understanding of the complexity of physical reality. By way of a response, in the final section of the paper I shall suggest that the interpretation of the instant and temporal discontinuity in relation to the void is given fresh impetus by Michel Serres' reading of Lucretian atomism as an articulation of the principles of non-linear dynamics.

1. Bachelard

> We cannot compare any process with the 'passage of time' – there is no such thing
> Ludwig Wittgenstein[4]

Bachelard presents his reading of time, as to a lesser extent does Levinas, in response to Bergson and it will therefore be helpful to begin with a brief outline of the perspective Bergson adopts on the question of time, with particular reference to nothingness or the void.

In Chapter 4 of *Creative Evolution*, Bergson describes two illusions that have beset the theoretical treatment of time. The first follows from the fact that through our intellect we grasp only instantaneous and thereby immobile views of the changing physical world. Perceiving only states, we succumb to the illusion that 'we can think the unstable by means of the stable, the moving by means of the immobile' [CE 288/Œ 726]. The second illusion is directly concerned with the way that we conceive of change in relation to nothingness or the void. If we are thinking of the void in physical terms, Bergson argues, we ordinarily mean the absence of a given thing in which we have an interest. Absence is treated as lack and thereby as different only in degree from being as plenitude. Moreover, as the lack of a specific thing, it is itself treated as local and defined, in the manner of a thing. There is no real interruption of physical reality. Similarly, if we are thinking of the void in terms of consciousness, I can interrupt the course of my thoughts, but must conceive of myself watching over this interruption and thereby filling the void again [CE 298/Œ 734]. At every turn, the idea of the void is illusory: however we seek to determine or conceive it, we find a continual and uninterrupted evolution. This is taken by Bergson as evidence against our common, cinematographic, approach to the apprehension of change, according to which the mind moves through a series of 'snapshots'. As Bergson explains, such an approach is bound to fail, since each time I concentrate on the transition between two states, a third slips in and I am left with an immobile image once more [CE 754–755/Œ 324]. The mind cannot grasp the transition without discovering new intermediary states. Bergson's conclusion is twofold. First, change is continuous: there are no breaks, no void. Second, the cinematographic model is inadequate to the task of grasping this continual change. In his own philosophy, he develops the idea of duration as a continuous and heterogeneous flux, and the concept of intuition as a direct entering into this flux.

Both *L'Intuition de l'instant* [1931] and *La Dialectique de la durée*, published five years later, develop their theses in direct opposition to the work of Bergson. This opposition on Bachelard's part is curious, insofar as they are apparently very close to Bergson in their respective aims and even in the terms they use. Above all, like Bergson, Bachelard is concerned that any thinking of time should be able to account for the occurrence of what is new. Bachelard, however, regards Bergson's approach as ultimately denying just what it would affirm. He concluded that to speak of time as essentially continuous is to allow that time passes independently of the events by which we perceive it. He therefore takes Bergson to be the latest expression of a traditional view whereby events are 'in' time in the sense that they are 'contained' by time. Such a view would entail that time persist between events, even when we are not aware of it, and even when it is apparently empty. The fundamental continuity of time would, he believes, ensure the underlying sameness or uniformity of time, its variety being derived from the 'secondary qualities' lent to it by phenomenal events occurring 'in' time. While this characterisation of Bergson's thought is quite contrary to Bergson's own interpretation of the movement of his philosophy, I will not enter the debate here, since I am more concerned with where Bachelard's opposition to Bergson leads him.

Recounting the failure of his early attempts to study Bergsonian duration directly, Bachelard writes that he was 'quite incapable of finding these endless, unbroken lines within us, the simple sweeping lines traced by the vital impetus in the picture it draws of becoming' [TI 77/II 33]. In spite of his efforts, he could not discover 'a life story in which nothing is lived, a happening in which nothing happens' [TI 77/II 32]. This search for a pure becoming was frustrated by what Bachelard tellingly calls 'the continual onslaught of all that is accidental and new' [TI 77/II 32]. Bachelard's pursuit of an element guaranteeing continuity drew him to the study of smaller and smaller fragments, but without success: no matter how sharp the focus of attention, duration would not resolve into either a still image or the pure flow of time as such. For Bachelard, our confidence in the natural continuity of time is the result of inattention [TI 65/II 15].[5] Actions, states, time itself, seem continuous only insofar as we satisfy ourselves with general trends, but the more attention we pay, the richer the detail that is revealed and the greater the discontinuity. Time comprises a multiplicity of discrete events and each instant is inflected differently to the last [TI 77/II 33].[6] For Bachelard, any general or formal determination of time 'as such' will capture at best this abstract continuity whose emptiness is the result simply of inattention. By contrast, time in its concreteness teems with a multiplicity of discrete events, unpredictably 'alive'.[7]

Reversing Bergson's thesis, then, Bachelard declares that it is continuity that is an abstraction having 'no absolute reality' and the instant that is real [TI 72/II 25]. He describes how he began by treating the instant as a fragment of Bergsonian duration, but soon realised that this hybrid account would not work. This led him to abandon what he calls temporal atomism, by which he means the idea that time could be built up from a collection of discrete instants. This idea of the instant as a unit of time is just what Aristotle, and every subsequent defence of the continuity of time (including that of Bergson), rejected as incoherent. To an extent, Bachelard may have shared this scepticism and his reticence to describe his approach to time as atomist reflects his decision to address the instant from an epistemological perspective rather than an ontological perspective. Nonetheless, atomist themes are clearly present. For example, on Bachelard's account, between events time is not just empty, there is simply *no time*: 'time is nothing if nothing happens in it' [TI 81/II 39]. Instants are separated by a temporal void: not a lull that can be measured or even imagined, but the absolute absence of time in precisely the sense that Bergson found inconceivable. The void here is not an ontological absence, but 'repose' punctuating instants as acts. It signifies discontinuity and is central to Bachelard's translation of ontological character into the temporal form of rhythm and frequency. Reality acquires shape and intensity in the instants through which things occur in their relations. As conscious beings, we do not just spectate as life goes on around us: 'understanding life involves not just living it but in fact propelling it forward' [TI 70/II 22]. Time and life are linked via the instant as an act. One clear implication of this is that the changes of which we become aware in instantaneous acts of attention receive their temporal character, their rhythm, from our own efforts at understanding them. While we are constantly assailed by chance events, these events are themselves inseparable from 'the rhythm of our acts of attention' [TI 79/II 35] (indeed it is precisely this inseparability that leaves us temperamentally exposed to chaotic events).[8] Through the instant we discover not just the *extent* of change, but also its *intensity*, the degree of detail and novelty it manifests. As we have noted already, the more we look, the more we see. And what we find, in Bachelard's view, is not only additional detail in the structure of phenomena, but also embedded within them degrees of chance and accident that previous levels of attention failed to register. It is the void, as the condition of the discreteness of the instant as an act, that gives a place in this account to chance – and thereby complexity – as an inherent feature of our reality, and not just as a failure on our part to fully understand that reality.[9]

The minimal condition for the occurrence of what is new is a coincidence of elements, by which we mean the occurrence of a relation, since no individual element has a continuous existence independently of its coincidence – its relation – with others. No individual thing is simply what it is independently of its context, its associations and adumbrations; and these are constantly shifting in ways that are both unpredictable and intermittent. With the instant, then, we are introduced to a time of chance; by which we understand not just the chance physical event, but also the non-causal coincidence of qualities, sensations and intensities brought together in our acts of attention. In spite of its apparent simplicity, the instant can therefore be described in terms of a complexity arising from the relational character of the instantaneous condensation of distinct elements into a single intensity – elements that may not exist independently of the instant in which they occur as phenomena. For this reason the instant cannot be treated either as an abstraction or as an entity in its own right. It must be treated precisely as a relation. Moreover, priority must be given to the relation itself and not the terms of that relation as independent relata.[10] This, as we shall see, is something that Bachelard's account struggles to accomplish. From a

The Complexity of the Instant

scientific perspective, Bachelard suggests that atoms may be considered actually to exist only at the instant of their interaction with other atoms. As he writes, between events the atom may be said to have 'an entirely virtual energetic existence' [TI 91/II 54]. In modern physics, individual changes in energy levels at the atomic and sub-atomic level cannot be determined causally and the precise moment and nature of the event is a matter of 'chance' [TI 91/II 55]. In this way, while the behaviour of the overall system can be treated mathematically via a statistical summation of probabilities, the individual atom or particle does not 'have' a time in which it can exist, radiate, interact. Its temporal existence is shaped rather by the possible interactions available to it and ultimately by the number of instants of which it disposes – and this cannot be reliably predicted or determined [TI 91/II 55].

However, this characterisation of the instant in terms of physical events becomes problematic when Bachelard describes his attempt to discover the most fundamental level at which the ocurrence of the new can be discovered. Breaking the rhythm of his attention down as finely as possible, he confronted the 'minimum of unexpectedness' not in a physical event, but in the realm of pure inwardness 'stripped of all else' [TI 79/II 35].

> We suddenly realised that this attention to ourselves brought with it, by virtue of the way it worked, all the fragile and delightful newness of thought that has no history, of thought without thoughts. [TI 79/II 35]

If the instant is to be new, to break with continuity, then it cannot have a history. It cannot be a vehicle for the causal influence of the past on the present and the future. But Bachelard's description of the instantaneous act as a 'thought without thought' is initially perplexing on two counts. First, if the instantaneous act is without content, in what does its newness consist (how does it differ from the last)? Second, is it not thereby treated as a 'pure' form of the instant as such, abstracted from any coincidence given in the act of attention? In this respect, how does it differ from the continuous but empty duration already criticised by Bachelard as 'a happening in which nothing happens [*une incidence sans incidents*]' [TI 77/II 32]?

Up to a point, these questions are easily resolved, insofar as they are *posed* from the point of view of continuous duration, presupposing that phenomena must appear 'in' time, which will remain empty in the absence of events. In this way, the first question assumes that newness must consist in the modification of the content to which we attend.

However, for Bachelard the instant of attention is new simply by virtue of its being given again, *another time, in another act*. Its content is irrelevant; or rather, its content will be itself different by virtue of the distinctness of the act in which it is apprehended. This already points towards a solution to the second question, in that the absence of thoughts in the thought is not the absence of content in a form that remains the same, but the concentration of the act upon itself as new each time. Its newness is not indebted to the memory of the previous instant. There is no temporal gap in which a thinking subject that already exists reflects upon its own activity – the focus is entirely on the act itself. Instantaneity is the condition both of putting the relation first, and of that relation being withdrawn from the linear time of cause and effect. As such, it is the condition of the occurrence of the new. However, precisely by virtue of the original inseparability of its terms, instantaneity also resists analysis. If the coincidence by virtue of which the act of attention occurs as a *new* instant is not to remain an absolute about which we can say nothing (beyond referring to it in the enigmatic terms of decision, will and vital impetus [TI 67, 70/II 18, 22]), we need to find a way to speak of the 'concealed' dimension in which it occurs.

In an essay entitled '*Instant poétique et instant métaphysique*' [1939],[11] Bachelard attributes the work of articulating coincidences to poetry. Building on Baudelaire's *correspondances*, he describes the way the poet breaks down the continuity of serial time in order to construct a complex nexus of simultaneities – an instant. Undoing the usual temporal order, the poem achieves an instant in the sense that it places together elements that had previously escaped our attention, or more especially that our usual 'prosodic' time does not allow us to see in their coincidence. Antitheses formed in serial time interest and please the poet, but a poetic instant is most properly composed only when they contract to form an ambivalence. More precisely, the poetic instant comprises a contradictory singularity in that it does not obey the law of non-contradiction, which requires that thesis and antithesis be linked by a single continuous time of negation. The poet lives *in an instant* the two terms of the antithesis without separating them logically, causally or temporally. The second term does not recall the first, or the first point on to the second: the terms are 'born together' [II 105]. There is an original multiplicity in which horizontal time is checked and held motionless [II 109]. In doing so, the poem discloses a complexity ordinarily concealed within attention as an instantaneous act: a sensible experience is composed, we see things a certain way. Poetry thereby presents the instant *as*

The Complexity of the Instant

complex, and in poetry we experience the complexity of the instantaneous [II 106–7].

Is this still time in any recognisable sense? Bachelard concludes that it is, insofar as the simultaneities or coincidences are still *ordered*, even though their relation cannot be composed in time as we ordinarily understand it. The order is a coincidence of elements that the usual conditions of sense insist on separating (logically and temporally). Yet the equilibrium within the instant is by no means inert or static and Bachelard describes it as 'an excited, active, dynamic ambivalence' [II 104]. This is to say that the 'order' is experienced as oscillating, fluctuating. 'Time no longer flows, it wells up [*jaillit*]' [II 106].[12] Bachelard calls this instantaneous time of the poem *vertical time* to distinguish it from the usual conception of time as flowing horizontally.[13]

In *The Dialectic of Duration,* Bachelard elaborates his idea of vertical time in terms of the superposition of different temporal sequences. Drawing on work in contemporary psychology (and that of Minkowski in particular) where patient's 'immanent time' was seen to have fallen out of step with the 'transitive' time of external reality, Bachelard records a dream of his own in which the time of speech fell out of synchrony with visual time. When the two orders are superimposed coherently one upon the other then the different mental faculties and processes combine to produce an impression of objectivity: 'We then say what we see; we think what we say: time is truly vertical and yet it also flows along its horizontal course...' [TDD 108/LDD 99]. Vertical time is presented here as the dimension of a coincidence between different orders of horizontal time. *The instant is the point where our reality coheres from the superimposition of differing temporal sequences and rhythms; not a synchrony within a common frame of reference, but rather the composition of a such frame.*

To examine the instant as the dimension of this vertical coincidence Bachelard brackets the time of the mind or the self from that 'of the world and of matter' [TDD 108/LDD 98]. Similarly, the material element of the self and the self's historical experience are also set aside in an extraordinary stratification of the Cartesian cogito. In place of 'I think therefore I am', Bachelard sets 'I think that I think, therefore I am', and even 'I think that I think that I think...' [TDD 108–109/LDD 99]. In this progressive formalisation of the self, the I is gradually distanced from being, and the discursive and still worldly time of the basic Cartesian cogito is deepened into 'the very first adumbration of vertical time' [TDD 109/LDD 99]. In looking at *L'intuition de l'instant* we queried the sense in which our focus on the instant of attention might bring us to 'the fragile and delightful newness of thought that has no history, of thought without thoughts' [TI 79/II 35]. We suggested then that it need not be read as the emptiness of an unchanging form with no content, explaining it rather as the coincidence generated by the act of attention turned upon itself. Here we read Bachelard describing the tiered cogito as the activity not of 'thinking *something*', but of 'thinking ourselves as *someone* who is thinking' [DD 109/LDD 99]. Again, we might usually speak of this in terms of reflection, whereupon there is a divide between the subject and the object of reflection. But as we have already seen in the case of the poetic instant and as we find again here, the coincidence indicated by Bachelard in the instantaneous act, in vertical time, is not sequential in this way. The act of 'reflection' is concentrated into a tautology that Bachelard calls a guarantee of instantaneity [TDD 110/LDD 100]. By this, Bachelard apparently means that we cannot fail to discover here at least a minimal newness – that of the act itself as a new venture or effort. Even in the Bachelardian 'I think that I think that I think…' the self, protected from the onslaught of chance events, still discovers an irreducible complexity within itself.

Here is at once the strength of Bachelard's account, and the point where it runs up against a limit. For the complexity Bachelard discovers in the instant is irreducible to the extent that the instant as relation is treated precisely as prior to the relata. But in *The Dialectic of Duration*, the vertical axis of time is explicitly associated with the time of the mind, and the horizontal axis of time with the time of life. If we began by taking the instant as the relation between consciousness and reality, we have ended by taking it as immanent to the mind's relation to itself. Is the act of the tiered cogito an exemplary case of the coincidence characteristic of the instant in general? If not, if the time of the mind is indeed fundamentally distinct from the time of life, then it seems that Bachelard situates the complexity of the instant in the vertical time of the mind. As such, all that we can learn from the idea of vertical time must at the very least be translated into another register before it can be relevant to complexity in physical reality. Given the separation between the time of mind and the time of life, it is not at all clear how such a translation could be carried out. Yet the dimension it would span is precisely the dimension of the instant in its priority as a relation. In spite of everything, Bachelard's account of the instant seems to retreat before this priority, which is addressed only indirectly through the idea of vertical time. This seems to be due above all to Bachelard's reliance on the discourse of psychology and more generally to his own epistemology, in which mind and reality are at once wholly distinct and locked into a relation of reciprocal modification.[14] While the point of their relation can be identified, it remains ultimately enigmatic. Yet to pursue the idea of the instant as the articulation of complexity, this division must be opened up and explored *as a relation*, and not as a conjunction of independent terms. Arguably, no discourse has gone further in this direction than phenomenology and certainly no one in the phenomenological tradition has explored the idea of the instant more fully than Levinas.

2. Levinas

> We must return from the world to life, which has already been betrayed by knowledge.
> Emmanuel Levinas[15]

In 1946 Levinas published *De L'existence à l'existant* in which the problematic of the instant plays a central part. The book's declared motivation is the need to 'leave the climate of Heideggerian philosophy' [EE 19/DEE19], but in doing so it sets itself against the whole philosophical tradition that has treated the instant as an abstraction from time [EE 74/DEE 126–127]. Indirectly, it is therefore aligned with Bachelard's analysis, from which it appears to have taken over several themes and concepts.[16] However, where these themes and concepts appear in Levinas, the context of their development is phenomenological rather than scientific and psychological.

Like Bachelard, Levinas takes the instant as a time of beginning that is quite new and not a continuation of what preceded it [EE 76/DEE 131].[17] The instant therefore revisits an old paradox in that 'What begins to be does not exist before having begun, and yet it is what does not exist that must through its own beginning give birth to itself, without coming from anywhere' [EE 76/DEE 130]. Far from seeking to avoid or resolve this paradox, Levinas underlines his desire to break with the paradigm of linear and continuous time by reinforcing it as a paradox, adding that 'A beginning does not start out of the instant that precedes the beginning; its point of departure is contained in its point of arrival, like a rebound movement [*un choc en retour*]' [EE 76/DEE 131].[18] Like Bachelard, Levinas associates this peculiar dynamism with the intensity of the self-relation dramatised in the Cartesian cogito. However, the apparent vitality of the instant in Bachelard's interpretation is counterbalanced in Levinas by a reading of the ego as bound to itself in the instant in such a way as to experience its irremissibility as an extreme burden. Moreover, Levinas also describes the instant in such a way as to lend the coincidence framed within it the pathos of distance: the paradox of an absolute beginning announces a sudden vulnerability to a dimension incommensurable with that of temporal continuity. The self comes to itself 'without having left from anywhere' and the essence of the instant is its spanning this 'inner distance' that cannot be presented [EE 77/DEE 131].

The excess of this temporal anomaly with respect to presentation is given a more fully developed phenomenological interpretation in the essay 'The Ruin of Representation' [1959], where it becomes central to Levinas' interpretation of the intentional structure of phenomenological consciousness studied and elaborated above all in the work of Husserl. Intentionality designates a relation of consciousness to its object in which the meaning of that object exceeds what is *explicitly* intended at any instant by virtue of the *implicit* horizons against which the object is apprehended. The intention is thereby characterised by excess, bearing within itself the 'innumerable horizons of its implications' and thinking 'of infinitely more 'things' than the object upon which it is fixed' [DEH 116/EDE 130]. Levinas determines that this excess cannot be formulated in terms of a continuous time in which the implicit and 'potentially' present implications were just by degrees 'less' present than the explicit object.[19] Rather, the implicit horizons are found to play a transcendental role in the constitution of the intentional object, even as they themselves are constituted by the intention on which their existence depends. In this way, everyday consciousness conceals within itself a relation whose paradoxical temporal structure is just that of the instant as Levinas described it above. This 'ambiguity of constitution' is played out within sensibility and sensuous experience which, as a consequence, is no longer structured by any prior transcendental forms.[20] Phenomena are not governed by universal physical law, logic or innate forms of consciousness that stand apart from sensuous experience itself. This in turn means that consciousness need not become 'consciousness in general'; it remains a 'unique life' whose intentionality is rooted in a 'living present' [DEH 141/EDE 152]. The adjective 'living' designates the coincidence whereby consciousness is at once consciousness of an object *and* a non-objectifying consciousness of itself [DEH 138/EDE 149]. But Levinas notes that in addition Husserl applies the term to sensations. More precisely, without the sensations by which the object is represented to consciousness, there could be no *act* of self-consciousness at all. The act can therefore never be just coincident with itself alone. The 'living present' is seeded by sensation.

The Complexity of the Instant

As Levinas writes, at each in a series of instants, the sensation is 'lived' via an intentional relation that is for Husserl 'just the very consciousness of time' [DEH 142/EDE 152].

> Time is not only the form that houses sensations and lures them into a becoming, it is the sensing of a sensation, which is not a simple coincidence of sensing and the sensed, but an intentionality and consequently a minimal distance between the sensing and the sensed – precisely a temporal distance. [DEH 142/EDE 152-3]

Here, the distance between sensing and sensed, however minimal, remains a temporal distance. The situation is more complex in the following passage.

> An accentuated, living, absolutely new instant – the proto-impression – already deviates [*s'écarte*] from that needlepoint where it matures absolutely *present*, and through this deviation [*écart*] presents *itself*, retained, to a new punctual present, sensed in advance in a protention departing from the first proto-impression, and including in this presentiment the imminence of its own retreat into the immediate past of retention. [DEH 142/EDE 153]

Although the temporal distance between sensing and sensed is described as the *advance* achieved in protention sensing the 'next' proto-impression *from* the first, and is to this extent still durational or horizontal, the first lines of the passage as it appears here suggest another reading. That the intentional structure is still necessarily accompanied by sensation means that the protention and retention by which flux is effected could not occur without the deviation [*écart*] of the proto-impression. But how does this deviation occur? While Levinas makes it perfectly clear that the deviation occurs in and through retention and protention, it is not simply effected by them. Rather, the proto-impression is at once *absolutely present* and *already deviating*. The force of Levinas' expression, here, lies in its placing together two terms that time demands we separate (the present and the past, indicated by adverb 'already'). As Levinas writes in the lines that immediately follow:

> Quite curiously, it is the present's acute and separated punctuality that constitutes its life; to it retention and protention, through which the lived flux is consciousness of time, attach their moorings. [DEH 142/EDE 153]

Retention and protention latch on to the present, which is punctual without being a simple fixed point. It is punctual in the sense that it is instantaneous, as we cannot measure or imagine a time between the instant's being *absolutely present* and its *already* deviating from that present. In spite of the movement enacted by retention and protention, the deviation is not reducible to a horizontal dimension: it does not occur 'in' time (otherwise we would be left with the mind conjuring a purely subjective time out of a series of inert presents, Bergson's illusory cinematographic time). It is the instant, in and through which time occurs.

Levinas refers to this deviation in two further ways: as a modification and as a de-phasing.[21] To speak of it as a modification, a term borrowed directly from Husserl, seems to hand too much back to continuity, given the classical account of change as the modification of accidents associated with a substantial identity. De-phasing is a more intriguing possibility, invoking pulses, frequencies and even a Bachelardian play of coincidence in vertical time. It is also the term that, recalling his early analyses of the temporal lag contained within the instant, Levinas uses when describing how the presence to each other of the world and the subject cannot be crystallized into a *structure* of simultaneity [DEH 148/EDE 159].

Nine years later, in the essay 'From Consciousness to Wakefulness' [1974], the idea of de-phasing occurs again at a crucial stage of Levinas' exploration of the 'living presence' of the ego to itself, described in terms borrowed once again from Husserl as 'transcendence in immanence' and as 'an incessant awakening' [DEH 161–3/DVI 49–52]. Once again, too, the role of sensation is decisive. Levinas begins by describing the awakening of the ego – its self-relation – as depending on the affection it undergoes in its 'collision with the real' [DEH 161/DVI 49]. Awakening here is characterised as a response to an alterity that the ego must assimilate in the accomplishment of its self-relation. This account is consistent with the standard account of intentionality given by Husserl in *Ideas I* that implicitly acknowledged the fact of intentionality prior to its engagement with sensation. Levinas then notes that in *Experience and Judgement* and *Phenomenological Psychology*, Husserl gives a different reading of intentionality, as coincident with the affected and awakened ego. Even in its most 'virtual' form, the ego is *'never torpid to the point of absence'* [DEH 162/DVI 49]. This lays to rest the possibility of conceiving the awakening of the ego – in the instant – as the encounter between sensation and a pure form of consciousness that lies passively 'waiting' to be brought to life. Now, having modestly elicited the most remarkable series of readings from Husserl's texts, Levinas avers that the time has come to push beyond their letter (if not their spirit).

Speculating on the life of consciousness before the emergence of intentionality as relation, before even the deviation of the proto-impression

The Complexity of the Instant 203

understood as effected by protention and retention, Levinas introduces the idea of insomnia, a condition of partial waking in which the ego is not fully present.[22]

> In the identity of the state of consciousness present to itself, in this silent tautology of the pre-reflexive, *wakes* a difference between the same and the same, never in phase [*jamais en phase*], which identity cannot clasp: precisely the *insomnia* that cannot be said otherwise than by these words, of *categorial* meaning. [DEH 162/DVI 50; translation modified]

Not a difference between the same and the other, but between the same and the same: a minimal difference or deviation, the slightest de-phasing that nonetheless resists the fixed structures of identity – though it cannot be articulated in any other terms. Not a bridge between non-being and being, sleeping and waking, inertia and life, nor between subject and object, nor between future and present or present and past, but a suspension that is not defined by its poles, neither one nor the other. The difference that awakens is not an *effect* of an external cause (as such, it is indeterminate, accidental). It is 'A waking [*veille*] without intentionality, but only awakened [*réveillée*] unceasingly from its very state of waking [*veille*]' [GCM 162/DVI 51], a minimal deviation that can no longer even be attributed to the ego's awareness of the proto-impression.

The initial paradox of the inner dimension of the instant has been elaborated in terms of the diachronic structure of phenomenological consciousness and this in turn emptied of any structure implying either the reception of experience by a consciousness that exists in advance, or conscious life as always already in relation to an intentional object. In this way the analysis of the instant is carried beyond the point where tradition has always discovered time, namely in experience, at the interface between mind and reality, consciousness and the world, subject and object. The instant, here, is not the articulation of a complexity composed of several elements, but the emergence of the relation between conscious life and the world in all its complexity. One might almost say that, in spite of all their differences, Levinas has accomplished Bachelard's intention to think the instant as a relation prior to its relata.

We may be surprised, then, to read in the very same passage a description of the de-phasing that is insomnia as 'An irreducible category of difference *at the heart* of the Same, that pierces the structure of being in animating or inspiring it' [DEH 162/DVI 50]. Not only does the juxtaposition of the categories of difference and the Same invite a dialectical interpretation that Levinas expressly wishes to avoid, the image of difference piercing being in such a way as to *animate* it and *inspire* it inevitably invokes a familiar conception of spirit breathing life into inert matter.[23] Not only is this theologically inspired conception of life outmoded – Levinas admits as much himself – and therefore an unreliable guide to thinking through the issues raised by the instant and time, but it seems at odds with the vocabulary of deviation and de-phasing in which Levinas has couched his analyses up to this point. While the vocabulary of animation and inspiration is ultimately theological in its origins, Levinas in fact takes it over from Husserl's analyses of time. However, it receives a very distinctive interpretation in his own work and we shall look at this before going on to develop a reading in terms of deviation. To do this, we must look first at Levinas' conception of the void in his analysis of time, waking and insomnia. We shall then be able to open up the possibility of a materialist – atomist – reading of the minimal deviation [*écart*] in the instant that is cognisant of Levinas' concerns, while proposing a different solution to them.

The idea of the void first appears in Levinas' writing in *Existence and Existents,* where he considers the act of negation carried to its furthest extreme. When consciousness attempts to imagine total nothingness, it cannot do so without spoiling the scene by its own intrusion, remaining in spite of itself 'as an operation, as the consciousness of that darkness' [EE 63/DEE 103].[24] Such consciousness cannot help but produce a residual stratum of being, unyielding and oppressive. Yet without any object upon which to focus, consciousness itself loses its resolution and dissolves into a 'bare presence'.[25] All that remains is the 'existential density of the void itself, devoid of all being, empty even of void' [EE 64/DEE 104]. This is what Levinas calls the *there is* [*il y a*].

As Levinas himself remarks, this approach still regards nothingness as 'the end and limit of being, as an ocean that beats up against it on all sides' [EE 64/DEE 105]. Even conceding Levinas' radicalisation, such an interpretation of the void is at the furthest reaches of an itinerary that begins with Parmenides' warning against the contradiction of thinking or speaking of non-being. The void, as presented in this guise by Levinas, is not the absence of being, but the most extreme articulation of the impossibility of speaking, writing or experiencing such an absence. Where classical ontology regarded this impossibility as proof of the incoherency of the void, Levinas takes it as evidence of the inexpungeable insistence of the void beneath intentional consciousness and experience. The incoherency of the void in the eyes of classical ontology is thereby given a positive re-evaluation by Levinas as the basis from which to begin

The Complexity of the Instant 205

thinking intentional consciousness, and thereby time, beyond the constraints of dialectical negation. However, as long as the void is characterised *solely* as a presence or existential density, it will be an idea that keeps faith with the Aristotelian tradition; the idea of existence in the smallest degree, the first and most inauspicious mark on the scale whose summit might be the pure actuality of divine being (thought thinking itself). Acknowledging the impossibility of thinking *through* being to non-being, of securing an *ontological* discontinuity that does not perpetuate some deeper continuity, Levinas looks to adopt a different approach: 'We must ask if 'nothingness', unthinkable as a limit of negation of being, is not possible as interval and interruption' [EE 64/DEE 105]. Levinas develops the idea of nothingness, the void, as interval and interruption in terms of the advent of a subject, or the participation of consciousness in the impersonal wakefulness [*veille*] through which the *there is* occurs [EE 65–6/DEE 109–110]. It is interesting to contrast his approach here to that of Bachelard, for whom consciousness exists only in its act and time itself is intermittent, made discontinuous not by empty time, but by no time at all. In Bachelard, the void is thereby given a temporal interpretation as the lacuna between instants. Levinas begins from a more traditional ontological interpretation of the impossibility of the void, whereupon consciousness emerges not as an interruption of the void, but as an interruption of the state of minimal existence. In a reversal of the more conventionally atomistic approach of Bachelard, Levinas will see consciousness itself as comprising the void by which being is interrupted. And he too will give a temporal reading of the void, though one quite different to that given by Bachelard.

As we have seen, Levinas develops his interpretation of the living present beyond the point where it is understood as born of an encounter between subject and object. But there remains an ambiguity regarding the involvement of affectivity in this event. In Levinas' earlier work, the void still has a primarily ontological sense as a minimal existence, matched by the impersonal and anonymous condition of pre-conscious insomnia. This cannot stir itself to life through its own efforts and it must be affected by an other in order to 'traverse time' and, through a 'miraculous fecundity', recommence anew and thereby become a subject [EE 93/DEE 159].[26] In the later essays, the condition of insomnia, the void, is described as a waking [*veiller*] that is in addition *already* an awakening [*éveil*] of the ego from insomnia as a result of a radical affectivity.[27]

'Already' here signifies not a simple order of events, but a diachronic temporal anomaly like that in the instant. At the level of discourse it can only appear as an ambiguity (the ambiguity of constitution). For if the void and awakening were either contemporaneous or sequential in time, the interval Levinas introduces would still be continuous with the ontological condition of minimal existence and the conscious ego would to all intents and purposes be the actualisation of a possibility already lying dormant in the 'density of the void'. Instead, the void or insomnia is *already* waking because the two phases are implicated with one another in a manner akin to that of the intentional object and its horizons: that is, as precisely *not* present together. The introduction of the diachronic structure is necessary here in order to transform the void from the limit case of minimal existence and the ontological counterpart of insomnia to the waking of consciousness through its being-affected by a radically (i.e. infinite) other. *Affectivity is now diachronic, within the instant*. This is affection by what is not present as a 'real' object in the sense that it does not enter into a common structure with consciousness and 'does not form a *whole* with that which it affects' [GCM 118/DVI 183]. The diachronic instant is an evanescence within time, breaking with the duration in which it is grasped [EE 73/DEE 124–5] and concealing a transcendence whose significance is in no way exhausted by its implication in the dialectic of time: not a transition or a linkage of any kind, but rather an interruption, an interval, an affection without touch [*tangence*] [GCM 118/DVI 184]. It is in this unbridgeable interval within the diachronic instant, that one finds Levinas' own interpretation of the void beyond its merely ontological conception. This is a second temporal interpretation of the void, not as the interval between acts (Bachelard), but as the diachronic rupture of being by what in his later writing Levinas calls 'otherwise than being'.[28]

Levinas' analyses of diachrony open the way for a fresh approach to temporal discontinuity as the occurrence of a *new* event that is not determined in advance by either transcendental or empirical conditions. Levinas recognizes that the occurrence of the new depends on the existence of a complexity that cannot be expressed in the familiar order of continuous time. He finds this complexity in the diachronic structure of the instant as the waking of consciousness itself through an affectivity that can no longer be attributed to an identifiable object, an affection without touch (as it must be if the instant is not once again to be abstracted from a continuous time of causal relations).

The determination of deviation specifically as diachronic reflects Levinas' conviction that *any* ontological thesis must inevitably be characterised by continuity, whereupon the preservation of the instant will

The Complexity of the Instant 207

depend on a radical transcendence otherwise than being. As we have seen, this amounts to an acceptance on Levinas' part that ontology will in the end always conform to the Parmenidean thesis that being is whole, continuous and without interruption even by non-being or the void in its usual guise. As a consequence, for Levinas the disruption of temporal continuity – nothingness as interval, interruption, diachrony – can only be an all or nothing affair; and so it is that we find transcendence in immanence, the vivacity of the living present, described as inspiration, animation, the irruption of infinity within the finite, and the tearing open of material reality by spirit.[29] Moreover, because it occurs as an *interruption* of the ontological order, this transcendence cannot appear as such at all; time and the instant are implicated with one another in such a way that the complexity of the instant is concealed. Levinas thereby saves the discontinuity of the instant at the expense of condemning time to simple linear continuity. On such a reading, the complexity expressed in the diachronic structure of the instant can have nothing to do with material reality. The instant in Levinas is ill suited to dealing with the complexity of time that arises from a complexity of the ontological order itself.

To be fair, this was not a problem that Levinas recognised as his own and he cannot be criticised for not responding to it. But the growing recognition of non-linear and complex processes as central to the phenomena studied in physics and biology has made the question of the temporal articulation of complex systems unavoidable. The phenomenological approach adopted by Levinas has helped us to avoid the difficulty of addressing the temporality of the relation of consciousness to reality from a psychologistic perspective. However, we now want to explore the possibility of a different characterisation of the 'minimal deviation' that Levinas identifies in the instant. In place of the diachronic relation with infinity, we shall consider the idea of a material fluctuation as understood from within the atomist tradition, with particular reference to the idea of the void.

3. Atomism: Touch and Deviation

> Like all philosophers passionately concerned with objective reality, Lucretius was a genius of touch, not vision ... The physics of Aphrodite is a science of caresses.
>
> Michel Serres[30]

Atomism names a doctrine first introduced by Leucippus in Greece in the 4th century b.c. and developed by his successor Democritus. It is widely held to have arisen as a response to the Eleatic philosophy of Parmenides and Melissus, according to which being, what exists, is uncreated and immutably one. In denying the existence of the void, this ontology ends by denying also the reality of all change and movement. Democritean atomism accepts the principle of the indivisibility of what is real, but applies it not to reality as a whole, but rather to individual atoms that are free then to move in the void and to combine with other atoms to form the large-scale structures that make up our familiar world. In this way, the basic principle of the immutability of what is real was reconciled with an account of phenomenal change.[31] Epicurus elaborated his own version of atomism, which in turn served as the basis for the great account of atomist physics, epistemology, linguistics and morality given by Lucretius in *De rerum natura*.[32] It is on this final work that I shall focus here, and in particular on the reading given of it by Michel Serres in *The Birth of Physics*. What distinguishes this reading is Serres' recognition of Lucretian atomism as more than just an interesting though now defunct precursor to later ideas. When allied to the mathematical treatises of Archimedes, Lucretian atomism reveals itself to be a rigorous elaboration of a philosophy built around non-linear dynamics.

The void in classical philosophy has been viewed either as empty space or as emptiness in the sense of a localised absence, an empty place.[33] Both interpretations have their merits and have textual support.[34] However, neither encourage due recognition of the importance of the void as a condition for the dynamic character of material reality as it is to be found in Lucretius. In *De rerum natura*, he describes how the origins of all things lie in the fall of atoms through the void like an infinite cosmic rain. However, the void is neither just this infinite emptiness through which atoms fall, nor the room into which atoms move as they collide with one another. Its additional significance lies in its interruption of the ontological and dynamic continuity that would otherwise obtain. As the atoms fall, their laminar flow is interrupted by spontaneous minimal deviations that occur sporadically and unpredictably, *incerto tempore, incertisque locis*.[35] This uncaused declination is the *clinamen*. It is the minimal angle of deviation from laminar flow required to cause turbulence, and it is through turbulence that vortices emerge that lead in turn to a temporarily settled order of combinations between atoms. This stability may persist for a time, as long as the combinations between atoms slow their passage sufficiently to keep them from rejoining the open flux. Structure, order and form are contingent and local, both in space and time. The clinamen is the principle expressing the aleatory and stochastic character

The Complexity of the Instant

of all change and thereby the emergence of all order, including that of life itself.

As a minimal angle of deviation, the clinamen may be treated as an infinitesimal. From this perspective, the fact that we cannot determine precisely when it occurs might be accounted for by its being smaller than the finest possible unit of measurement: it thereby appears instantaneous simply by virtue of its being too brief to measure. The problem with this interpretation is that it presents the instant as a fragment of continuous time, different from duration only in degree. But there is another possible reading of the temporal indeterminacy of the clinamen. If the clinamen here is the event at which the path of the atom falling deviates from a straight line, it could only be in principle fully determinable when described in the idealised geometrical terms of infinitely thin lines and the like. Thus, if the instant is the time at which this event occurs, it too will be indeterminable. It may be, therefore, that the instant at which the clinamen occurs cannot be specified because it is intrinsically unstable, indeterminable, complex.

In view of its significance in Levinas' treatment of affectivity in relation to time, it is interesting to consider the role played here by touch. In fact, touch becomes complex in atomism, as we can no longer rely on the simplicity of ideal geometrical elements in deciding, for example, when two lines or surfaces are in direct contact with one another. For example, Serres describes the method of exhaustion in which a polygon is drawn inside a circle and the number of its sides increased so that its form approaches that of the circle.

> Now this operation, strictly speaking, both has and does not have an end. ... Now observe the fluctuating muddle that separates and unifies the border and the conjunction, the limit surface and the infinite increase in angles. Literally and without metaphor, this space is fluent. [BP 102/ NP 129]

The circle and the polygon are never simply coincident. Because their surfaces are not ideal, perfectly smooth and infinitely thin lamina, the space between them is irregular and lacunary, interspersed by the void.[36] Similarly, as the minimal deviation occurs, the curve departs from its tangent and one path becomes two – yet one cannot say precisely when there is just one line, when there are two lines touching and when the two lines are separated from one another. To describe this space as 'fluent' is to say that it is both indeterminate and indeterminable. As a consequence, touch cannot be the simple mechanism of cause and effect.

Let us see what happens when we take these conceptions of deviation and touch and read them back in to the account of minimal deviation that we have already seen in Levinas. In the essay 'Intentionality and Sensation' he describes time not as the form that houses sensations, but as the very sensing of sensation [DEH 142/EDE 153]. The intentional structure of experience means that there is a 'minimal distance' between the sensing and the sensed that Levinas follows Husserl in defining as a temporal distance, a 'living, absolutely new instant' [DEH 142/EDE 153]. This instant or proto-impression 'already deviates from the needle-point where it matures absolutely *present*', and through this deviation presents itself to another proto-impression [DEH 142/EDE 153].[37] In this way, the divergent instants are linked together by the protentions and retentions of consciousness without being objectified. Levinas can thereby claim that consciousness of time *is* time [DEH 143/EDE 153].

Even so, to speak of time in terms of the 'sensing of sensation' continues to address the *relation* that is time as a 'black box' between consciousness and reality – something that both Bachelard and Levinas wished to avoid. Levinas' response was to drop reference to sensation altogether, to press beyond the condition of full consciousness and to describe the 'waking' that already stirs within consciousness independently of the intentional relation to sensations and the protentions and retentions of the proto-impression. At this point he abandons the language of deviation [*écart*], presumably because his earlier account understood it to require a cause or touch to bring it about. Instead, he writes, as we have seen, of a minimal difference between the same and the same, 'never in phase'. This is a de-phasing without cause, without even the mechanism of protentions and retentions to carry it through. It is here that Levinas appeals to the idea of affection without touch, introducing a radical transcendence. Strangely, the 'minimal' difference between the same and the same is thereby cast as an infinite and unbridgeable separation – diachrony, the ungraspable irruption of infinity within finitude.[38]

Yet the spontaneous occurrence of a minimal deviation without cause is precisely what is theorised in Lucretian atomism as the clinamen. Minimal deviation can be presented without relying on the *touch* of retention and protention – which for Levinas tied time back into causality and what he perceived as the continuity of the ontological order. Where Levinas went beyond the active role of sensation, he appealed to the idea of infinity and a radical affection of consciousness by what is otherwise than being. By contrast, atomism can continue to focus on the *relation* that is the instant and can at least go further than either Bachelard or Levinas

The Complexity of the Instant

in describing it in its own terms. Moreover, it can do so without presenting an account of the instant 'as such', or reducing the complexity inherent within it.

Via a material discourse, it avoids any appeal to theological terms and to affection without touch. Because for atomism touch does not equate to causality and moreover because it is not itself instantaneous, as the geometrical model would have it, the instant of this touch or affection is not thereby placed beyond further analysis. *The atomist account allows, indeed invites, an exploration of touch as complex.* This means that the advantage gained by approaching the question of the instant and complexity via phenomenology – namely, that the account of complexity is properly focussed *in* the relation of consciousness to reality, and not split by the subject/object divide as in Bachelard – can be maintained, while avoiding the turn by which complexity in the instant is cashed out in terms of diachrony as a relation to infinity. From an atomist perspective, the complexity of the instant is the material articulation of complexity in the ontological order. In the end, this reformulates a view found in Bachelard, for whom phenomena are constructed by 'rhythms that are by no means necessarily grounded on an entirely uniform and regular time' [TDD 20/LDD ix].

Department of Philosophy
Staffordshire University,
Stoke-on-Trent, ST4 2XW, UK

Notes

1. The following abbreviations are used in this essay:
   BP   Michel Serres, *The Birth of Physics*, ed. D. Webb, trans. J. Hawkes (Manchester: Clinamen Press, 2000)
   CE   Henri Bergson, *Creative Evolution*, trans A. Mitchell (London: Macmillan, 1910)
   DVI   Emmanuel Levinas, *De Dieu qui vient à l'idée* (Paris: Librarie Philosophique, J. Vrin, 1992)
   DEE   Emmanuel Levinas, *De l'Existence à l'existant* (Paris: J. Vrin, 1963)
   DEH   Emmanuel Levinas, *Discovering Existence with Husserl*, trans. R.A. Cohen & M.B. Smith (Evanston: Northwestern University Press, 1998)
   EDE   Emmanuel Levinas, *En Découvrant l'existence avec Husserl et Heidegger*, (Paris: Librarie Philosophique, J. Vrin, 1994)
   EE   Emmanuel Levinas, *From Existence to Existents*, trans. A. Lingis, (The

Hague: Martinus Nijhoff, 1978)
GCM Emmanuel Levinas, *Of God Who Comes to Mind*, trans. B. Bergo (California: Stanford University Press, 1998)
II Gaston Bachelard, *L'Intuition de l'instant* (Paris: Stock, 1931)
LDD Gaston Bachelard, *La dialectique de la durée* (Paris: Presses Universitaires de France, 1993)
NP Michel Serres, *La naissance de la physique dans le texte de Lucrèce* (Paris: Les Éditions de Minuit, 1977)
Œ Henri Bergson, *Œuvres*, ed. A. Robinet (Paris: Presses Universitaires de France, 1959)
TDD Gaston Bachelard, *The Dialectic of Duration*, trans. M. McAllester Jones (Manchester: Clinamen Press, 1999)
TI Gaston Bachelard, *The Instant*, trans. M. McAllester Jones; Chapter 4 of this volume.

2 II 86.
3 BP 76/NP 97.
4 Ludwig Wittgenstein, *Tractatus Logico-Philosophicus*, trans. D.F. Pears & B.F. McGuinness (London: Routledge & Kegan Paul, 1922, 1961), § 6.3611.
5 Cf. also 'Saying that an action *has duration* means that we are still refusing to describe its detail' [TDD 38/LDD 19].
6 Again, this is very close to Bergson, for whom continuous duration was a heterogeneous multiplicity in constant variation (for an extended discussion of the possibility of sustaining a reading of duration as heterogeneous multiplicity, see R. Durie, 'Splitting Time: Bergson's Philosophical Legacy', *Philosophy Today*, 44 (2000), pp. 152–168). Bachelard is simply in disagreement over the conceptual means by which multiplicity can be addressed: He rejects 'the metaphysical extrapolation that argues for a *continuum in itself* when we are always faced with only the discontinuity of our experience' [TI 83/II 42].
7 The whole sentence runs: 'There was far more to duration than just having duration, for duration was truly alive!' [TI 77/II 33]. The reference to life here in relation to time is more than a casual use of metaphor on Bachelard's part. Indeed, with Roupnel, he explicitly affirms an absolute identity between the feeling of the present and that of life [TI 69/II 20]. No doubt Bergson's *elan vital* was at the forefront of his mind, and we shall see later that the idea of the living present is commonly used by Husserl and by Levinas. However, to speak of the present as 'living' is not merely an allusion to its being new. It is also to address time via an implicit appeal to a particular conception of life that has remained for the most part uninterrogated in this connection. Philosophy has perhaps remained more sheltered from findings in biology than from other sciences. For example, whereas the significance of the abandonment of absolute space has long since been appre-

The Complexity of the Instant 213

ciated in areas of philosophy not directly engaged with physics and mathematics, it is by comparison quite recently that repercussions of evolutionary theory have been widely felt. Similarly, the classical paradigm of life as the spontaneous unfolding of a potentiality into its actualized form has continued to leave traces long after its abandonment by those directly concerned with the issue in modern science. Even the theological idea of spirit animating lifeless earth still exerts an influence. These links cannot be passed off as secondary in thinking of time, as the two problematics – of time and of life – are related and findings in one will bear upon the other.

8   'Should we not agree then that it is metaphysically more prudent to equate time with what is accidental, which would mean in fact that time is to be equated with its phenomenon?' [TI 77/II 33] For Bachelard, the time of accident is at once the time in which we are assailed by events and the time in which we survive this assault through the formation of regularities. One moulds oneself from chance through the formation of habits. Cf. *L'Intuition de l'instant*, chapter 2.

9   'There is only one law in truly creative evolution, the law that an accident is at the root of any attempt at evolving' [TI 71/II 24].

10 Bachelard's adoption of this perspective on the priority of relation was prompted by his encounter with Einstein's theory of relativity. This point is made by M. McAllester Jones in her book *Gaston Bachelard, Subversive Humanist* (Madison: Wisconsin University Press, 1991), p.28. Bachelard discusses Einstein in *La Valeur inductive de la relativité* (Paris: Librarie Philosophique J. Vrin, 1929).

11 In *L'Intuition de l'instant*.

12 Bergson also uses the same word with respect to duration, describing the movement of the present as an 'upsurge [*jaillissement*]... in two jets, symmetrical, one of which falls back [*retombe*] towards the past whilst the other springs forward [*s'élance*] towards the future.' (Bergson, 'The Memory of the Present', TI p. 48)

13 '*En tout vrai poème, on peut alors trouver les éléments d'un temps qui ne suit pas la mesure, d'un temps que nous appellerons vertical pour le distinguer du temps commun qui fuit horizontalment avec l'eau du fleuve, avec le vent qui passé.*' [II 104] The idea of vertical time is also discussed in *The Dialectic of Duration*, where Bachelard refers to work in contemporary psychology by Marc, Straus, Gebsattel and in particular Minkowski [TDD 104–112/LDD 94–103].

14 '[T]he contours of the object are modified by the knowledge that draws these contours, and the criteria of precise knowledge are dependent upon the object's order of magnitude, the stability of its appearance, upon, as it were, its order of existence.' Bachelard, *Essai sur la connaisance approchée* (Paris: Librarie Philosophique J Vrin, 1928), cited in M. McAllester Jones, *Gaston Bachelard, Subversive Humanist*, p.20.

15 Levinas, 'La Philosophie et l'éveil' in *Études philosophiques*, no.3 [DEH 174].
16 Notably, the concepts of the instant, beginning, effort, habit. Many of these also appear in literature that would have been known to both of them, such as that of the psychologist Pierre Janet to which Bachelard refers in *The Dialectic of Duration* [TDD 55–57/LDD 38–40]. The concepts of effort and fatigue also appear in the following work: *Existence and Existents*, §2 'Fatigue and the Instant'.
17 See also 'effort' in 'Fatigue and the Instant'; and Bachelard, TDD 55–57/LDD 38–40. Just as significantly, Levinas also refers (in *Otherwise than Being, or Beyond Essence*, trans. A. Lingis (The Hague: Martinus Nijhoff, 1981), p. 33 [*Autrement qu'être ou au-delà de l'essence* (The Hague: Martinus Nijhoff, 1974), p. 41]) to Appendix I of *On the Phenomenology of the Consciousness of Internal-Time (1893–1917)*, trans. J.B. Brough (Dordrecht: Kluwer, 1991); *Zur Phänomenologie des inneren Zeitbewusstseins (1893–1917)*, ed. R. Boehm (*Husserliana X*; The Hague: Martinus Nijhoff, 1966), where Husserl writes: 'Modification continuously generates ever new modification. The primal impression is the absolute beginning of this production, the primal source, that from which everything else is continuously produced. But it itself is not produced; it does not arise as something produced but through *genesis spontanea*; it is primal generation. It does not spring from anything (it has no seed); it is primal creation.'
18 As in Bachelard's analyses, the instant is irreducible either to a time of potentiality and actuality, or to a causal order.
19 'To exceed the intention in the intention itself, to think more than one thinks, would be an absurdity if this exceeding of thought by thought were a movement of the same nature as that of representation, if the 'potential' were only a diminished or slack [*relâché*] form of the 'actual' (or it would be the banality of degrees of consciousness)' [DEH 117/EDE 131].
20 In this way, the relation to the transcendental is necessarily played out through sensibility, which is revealed not as the site merely of a particular set of problems (relating to perception etc.) but as the milieu from which *all* problems relating to consciousness proceed and *in which they must be addressed*. As Levinas writes, 'Sensuous experience is privileged, because within it that ambiguity of constitution, whereby the noema conditions and shelters the noesis that constitutes it, is played out.' [DEH 119/EDE 134]
21 For a comprehensive account of the way in which Levinas derives his understanding of the notions of modification and, in particular, de-phasure from a reading of Husserl's phenomenology of the consciousness of inner-time, see R. Durie, 'Speaking of Time... Husserl and Levinas on the Saying of Time', in *The Journal of the British Society for Phenomenology*, vol. 30.1 (1999), pp. 35ff.
22 Levinas first wrote of insomnia in *From Existence to Existents*, where it is

The Complexity of the Instant 215

characterized as an impersonal vigilance without any distinct object. It is developed there in conjunction with the *il y a* or 'there is', which is discussed below.

23 In 'God and Philosophy' Levinas provides a similar analysis of the soul as undergoing the 'passivity of Inspiration' [GCM 59/DVI 99], prefacing it on this occasion with the remark: 'to use an antiquated language'.
24 Cf. EE 64–65/DEE 103–5, where Levinas refers to Bergson's discussion of nothingness in *Creative Evolution* Chap. 4. Levinas effectively applies the Heideggerean idea of ontological difference to Bergson's reading: the impossibility of absolute negation leads not to the inescapability of a determinate being, but the inescapability of impersonal existence, an existence without a subject.
25 'The presence which arises behind nothingness is neither *a being*, nor consciousness functioning in a void, but the universal fact of the there is, which encompasses things and consciousness.' [EE 65/DEE 109]
26 'If time is not the illusion of a movement, pawing the ground, then the absolute alterity of another instant cannot be found in the subject who is definitively *himself*. This alterity comes to me only from the other' [EE 93/DEE 160]. Cf. also Levinas, *Time and the Other*, trans R.A. Cohen (Pittsburgh: Duquesne University Press, 1987), p. 79; *Le temps et l'autre* (Presses Universitaires de France, Paris, 1979), pp. 68–9.
27 On the relation between waking [*veiller*] and awakening [*éveil*] see J. Llewelyn, *The Genealogy of Ethics: Emmanuel Levinas* (Routledge, London, 1995) pp. 52–55.
28 Cf. Levinas, *Otherwise than Being, or Beyond Essence*.
29 Similarly, his formulations of this transcendence appeal to the rupture of the container by the uncontainable, of finitude by infinity; cf. 'Philosophy and Awakening' [DEH 178].
30 BP 134/NP 107
31 Cf. also C.Bailey, *The Greek Atomists and Epicurus* (Oxford: Oxford University Press, 1928).
32 There are several excellent editions of Lucretius in print at present, such as Lucretius, *On the Nature of the Universe*, trans. R. Melville (Oxford: Oxford University Press, 1999).
33 This formulation is drawn from D. Sedley 'Two Concepts of Vacuum' in *Phronesis* Vol. XXVII, No. 2 (1982), pp.175–193.
34 In the case of Epicurus and Lucretius, Sedley argues for an interpretation of void as empty place, not least because of the extent to which their ideas would have been formulated in the wake of Aristotle's rebuttal of the void (*Physics* IV, Chaps. 6–9).
35 The full phrase runs: 'While atoms move by their own weight straight down / Through the empty void, at quite uncertain times / And uncertain places they swerve slightly from their course. / You might call it no more than the

merest change of motion.' [trans. modified; *On the Nature of the Universe*, II 218–219]. Lucretius specifies that the deviation is no more than the minimum (*nec plus quam minimum*) [*On the Nature of the Universe*, II 244].

36 Cf. Bachelard's reference to time as lacunary [TDD 19/LDD vii].
37 The image of disequilibrium, and of time itself persisting through perpetual disequilibium, is a typically atomistic one.
38 Insofar as 'consciousness of the permanence of the flux ... is an effectuation of the permanence of the flux', Levinas can justifiably claim that 'Event and consciousness are on the same level' [DEH 143/EDE 153]. However, it seems that this cannot be said of the instant.

10
Time, Dynamics and Chaos

Dean J. Driebe

The irreversible passage of time is our most conspicuous existential feature. One would think that such a fundamental aspect of nature would have long been embedded in the framework of our scientific description of the world, but this is not the case. As traditionally formulated, the basic dynamical laws of physics, be they classical or quantum, are time-reversible. This means that time plays only the role of a parameter and can just as well be run forwards or backwards. There is no privileged direction of time in a classical mechanics or quantum mechanics description of nature.

This is certainly at odds with our experience of the world and leads one to either of two possible resolutions of this apparent antinomy. One viewpoint is to ascribe the appearance of irreversibility and a time direction to phenomenology. In this approach to the problem an explanation of the apparent breaking of time symmetry is sought in the way we observe nature. In some way we are responsible for the appearance of irreversibility in a world that is really timeless and the resolution of the problem consists in explaining how this appearance arises.

Another approach to the problem, advocated strongly by Prigogine, is to extend the formulations of both classical and quantum mechanics to include irreversibility.[1] In this approach, the traditional formulations of mechanics are seen as a framework applicable to idealized situations not present in the real universe. The direction of time is a property of our world and should appear in a proper formulation of the basic laws of nature.

The modern enchantment with a timeless view of the world emerged from the astounding success of Newtonian mechanics late in the 17th century. Newton's second law of motion along with his universal law of gravitation provided the explanation of, and more importantly enabled the prediction of, a vast range of natural phenomena. These included the motions of the moon and the planets, the explanation of Galileo's law of falling bodies and the phenomenon of the tides in their correlation to the positions of the moon.

The laws of motion formulated by Newton deal with the changes in velocity that a system of point particles undergoes due to the application of forces on the system. The dynamical problem is stated in terms of a set of differential equations with initial conditions and the solution to the problem, which corresponds to an integration of the equations of motion, gives the trajectory for the system. The trajectory is a curve in the abstract space of the dynamical variables. Once an initial condition, which entails the specification of the positions and momenta of all particles, is given, all future or past time development of the system is completely given by the trajectory solution. Newton's formulation furnishes a deterministic and time-reversible description of dynamical motion. On the other hand, the integration of the equations of motion is far from trivial and in general is in fact impossible. Such non-integrable systems include the simple problem of three particles in motion under mutual gravitational interaction.[2] Non-integrability of a system is associated with a complexity of the dynamics that has only recently become widely appreciated through the study of unstable dynamical systems and chaos.

The arrow of time entered into physics with the second law of thermodynamics. There are several equivalent formulations of this law, but in contrast to the first law of thermodynamics (which is simply the statement that energy is conserved, taking into account that heat is a form of energy) the second law is expressed as an inequality that places a limitation on processes. The law can be expressed in terms of a thermodynamic quantity, entropy, which must always increase in time for any process, except for those reversible processes where it would stay constant. But, as all thermodynamics textbooks state: 'The reversible process is an abstraction, an idealization, which is never achieved in practice.'[3]

William Thompson gave the first formulation of the second law of thermodynamics in 1852. He expressed it, as it still is today in many textbooks, as the existence in nature of a universal tendency toward the degradation of mechanical energy. This leads to the view of the universe as running down over time to an eventual heat death. The flow of time from this perspective has largely negative connotations. As expressed by Prigogine and Stengers:

It is generally accepted that the problem of time took on a new importance during the nineteenth century. Indeed, the essential role of time began to be noticed in all fields – in geology, in biology, in language, as well as in the study of human social evolution and ethics. But it is interesting that the specific form in which time was introduced in physics, as a tendency toward homogeneity and death, reminds us more of ancient mythological and religious archetypes than of the progressive complexification and diversification described by biology and the social sciences.[4]

Thermodynamics is a phenomenological theory in that it deals with the bulk, macroscopically observed and measured properties of systems consisting of many constituent elements. It was developed to deal with equilibrium states of systems. The state of the system is described with thermodynamic variables such as temperature, entropy and pressure. In an equilibrium state the variables remain constant. Equilibrium states are stationary states and so are timeless. Even the usual description of the change from one stationary state to another is described as a process where the system passes through a succession of stationary states.

The extension of thermodynamics to near equilibrium situations includes the consideration of extremal principles that account for the stability of equilibrium states. The usual thermodynamic variables may be used in near equilibrium situations and thermodynamic potentials must be extremized, driving the system back to equilibrium. But far from equilibrium these extremal principles are not valid and fluctuations may produce new and interesting space-time structures known as dissipative structures.[5] These include such phenomena as fluid instabilities, chemical oscillations and Turing patterns. This demonstrates the constructive role of far-from-equilibrium processes and shows that the naïve picture of nonequilibrium states as transients that will quickly approach equilibrium is in general not warranted. We recognize the results of far-from-equilibrium thermodynamics to be in accordance with the diversification that we see in nature. This tells us that we live in a far-from-equilibrium world.

Thermodynamic variables are macroscopic variables. The bridge between the microscopic dynamical aspects of the constituent parts of macroscopic systems and the thermodynamic variables is provided by statistical mechanics. The aim of statistical mechanics is to explain the thermodynamic properties of a system in terms of the behaviour of its constituents.[6]

The Austrian physicist Ludwig Boltzmann devoted much of his scientific career to a statistical mechanical explanation of the second law of thermodynamics. For dilute gases he derived an evolution equation

(Boltzmann's kinetic equation, written down in 1872) for velocities and the distribution in space of a representative particle in such a system and showed that an entropy describing an irreversible approach to equilibrium is obtained. His results were sharply criticized because the full distribution of all the particles in such a system inherits the time-reversible properties of the underlying Newtonian motion of the constituent particles so that the results seemed to contain a logical flaw. In fact, Boltzmann introduced irreversibility into his equation through an assumption known as molecular chaos. In his derivation, the evolution of the gas is driven by collisions of pairs of particles and he assumes that all such pair collisions involve uncorrelated particles. Since the collisions produce correlations, this amounts to assuming a privileged direction of time in the system. Even though Boltzmann's equation produces many beautiful results verified extensively by experiment, it cannot be accepted as a fundamental derivation of irreversible behaviour from time-reversible constituent dynamics. And the extension of Boltzmann's equation beyond the regime of dilute gases is not possible from his derivation and must be done using a different approach.

One of Boltzmann's sharpest critics was his contemporary, the French mathematical physicist Henri Poincaré. The work of Poincaré on dynamical systems, especially the problem of three bodies in motion due to mutual gravitational interactions, is seen now as the precursor to the modern theory of unstable or chaotic dynamical systems. Ironically, this field has revitalized[7] the program begun by Boltzmann and shows us that time and dynamical instability are deeply connected, as Prigogine has emphasized for some time now.[8]

One of the key ideas is that for unstable dynamics the concept of the trajectory loses operational meaning and must instead be replaced by a probability distribution. When the basic dynamical laws are expressed on this level there appear new solutions to the evolution equations that explicitly break time symmetry.[9] These new solutions are irreducible to trajectories and show us that a probabilistic formulation of dynamics is time oriented even for systems of few degrees of freedom.

An interesting feature of this new formulation is that there appears a link between the initial conditions and the dynamical evolution of systems. This link refers to solutions of the dynamical evolution problem which are different for classes of initial conditions that may be reduced to trajectories and for those that cannot. This difference is most prominent for systems treated in the so-called thermodynamic limit. This is a calculational procedure used in statistical mechanics to deal with large

Time, Dynamics and Chaos

systems that are described with both extensive and intensive variables. Extensive variables, like mass, depend on the size of the system, whereas intensive variables, like temperature do not. A description where there are intensive variables compels the use of probability densities that are irreducible to trajectories. Then solutions with broken time symmetry appear naturally already on the level of time evolution of the exact probability density of the complete system.[10]

The evolution equation for the full probability density is known as Liouville's equation. There are both classical and quantum versions of this equation that are derived directly from the equations of motion of the constituent particles, either Newton's equation or Schrödinger's equation. It has always been assumed that this equation is equivalent to the underlying equations of motion, thus also giving a time symmetric evolution for the probability density. As a recent textbook states:

> All averages such as position and velocity can be rigorously calculated from the Liouville equation. However, this equation is generally difficult to formulate and to solve. The solutions of the Liouville equation are reversible, so that the equation must be modified in some way if it is to specifically show irreversibility.[11]

This is a very strange, but surprisingly prevalent, viewpoint on the foundations of non-equilibrium statistical mechanics and the origin of the arrow of time. Physicists are so reluctant to give up an absolute time reversibility of dynamics that they prefer to build a whole subject on an apparent contradiction. The Liouville equation is the basic equation of statistical mechanics, as Newton's equation is for classical mechanics and Schrödinger's equation is for quantum mechanics. But in order to obtain a realization from this equation of the primary observational fact of irreversibility it is claimed that the equation must somehow be changed!

The way out of this predicament is, as stated above, the realization that there are solutions to the reversible Liouville equation that are themselves nevertheless irreversible. These solutions involve a somewhat unconventional mathematical setting (generalized function spaces) for the Liouville operator but recover all conventional results as well as the naturalness of a broken time symmetric formulation.[12] The Liouville equation does not need to be modified specifically to show irreversibility.

The construction of these new solutions requires a system with sufficiently complex dynamics. One class of such systems are those considered in deterministic chaos. Chaotic systems are characterized by exponential sensitivity to initial conditions on the trajectory level and by a microstructure of phase space on the statistical level. This microstructure means in fact that trajectories in such systems are invalid idealizations. Surrounding any initial condition with irregular behaviour appears another initial condition with periodic behaviour. Unless exact initial conditions are given not even qualitative features of the subsequent time evolution can be determined. Thus the introduction of a probability density sampling a finite region of phase space is warranted for a robust dynamical description. It is when we consider the time evolution of such non-local densities that irreversible solutions to the dynamical equations occur.[13]

The other class of systems for which such constructions have been carried out are those that are non-integrable in the sense of Poincaré.[14] This non-integrability refers to the failure of conventional perturbation techniques to obtain solutions to the dynamical problem. New techniques lead to solutions in generalized function spaces and irreducible irreversible representations of the dynamical evolution.

The construction of a formalism that allows us to see irreversibility on the most fundamental dynamical level gives us a unified conception of nature. This conception views probabilistic descriptions of dynamics as fundamental, even in a classical mechanics context. A probabilistic approach allows for time to play a real role in nature as it frees us from the rigidity of a deterministic trajectory description where the future is given.[15] The future is not given but is open to possibilities. This is a natural view of our human experience. Now we see that our experience is not alien to nature but is embedded in it. The flow of time is a real, objective property of our physical world.

Ilya Prigogine Center for Studies in Statistical
Mechanics and Complex Systems
The University of Texas at Austin
Austin, Texas, 78712, USA

Notes

1 Ilya Prigogine, *From Being to Becoming* (New York: Freeman, 1980); Ilya Prigogine & Isabelle Stengers, *Order out of Chaos* (New York: Bantam, 1984); Ilya Prigogine, *The End of Certainty* (New York: The Free Press, 1997).

2 F. Diacu & P. Holmes, *Celestial Encounters: The Origins of Chaos and Stability* (Princeton: Princeton University Press, 1996).

3   M.M. Abbott & H.C. Van Ness, *Thermodynamics* (Schaum's Outline Series; New York: McGraw Hill, 1972), p. 10.
4   Prigogine & Stengers, *Order out of Chaos*, p. 116.
5   D.K. Kondepudi & I. Prigogine, *Modern Thermodynamics: From Heat Engines To Dissipative Structures* (New York: Wiley, 1998).
6   Radu Balescu, *Equilibrium and Non-equilibrium Statistical Mechanics* (New York: Wiley, 1975).
7   See P. Gaspard, *Chaos, Scattering and Statistical Mechanics* (Cambridge: Cambridge University Press, 1998); J.R. Dorfman, *An Introduction to Chaos in Non-equilibrium Statistical Mechanics* (Cambridge: Cambridge University Press, 1999); W.G. Hoover, *Time Reversibility, Computer Simulation and Chaos* (Singapore: World Scientific, 1999).
8   Prigogine, *From Being to Becoming*.
9   Dean J. Driebe, *Fully Chaotic Maps and Broken Time Symmetry* (Dordrecht: Kluwer Academic Publishers, 1999).
10 T. Petrosky & I. Prigogine, 'Poincaré's resonances and the extension of classical dynamics', *Chaos, Solitons & Fractals*, vol. 7 (1996), 441–497.
11 R.E. Wilde & S. Singh, *Statistical Mechanics: Fundamentals And Modern Applications* (New York: Wiley, 1998), p. 232.
12 T. Petrosky & I. Prigogine, 'The Liouville space extension of quantum mechanics', *Advances in Chemical Physics*, vol. 99 (1997), 1–120; Petrosky & Prigogine, 'Poincaré's resonances and the extension of classical dynamics', Driebe, *Fully Chaotic Maps and Broken Time Symmetry*.
13 Driebe, *Fully Chaotic Maps and Broken Time Symmetry*.
14 Petrosky & Prigogine, 'Poincaré's resonances and the extension of classical dynamics'; 'The Liouville space extension of quantum mechanics'.
15 Karl R. Popper, *The Open Universe: An Argument For Indeterminism* (London: Routledge, 1982).

11
Time-shelters: an essay in the poetics of time

David Wood

> And into that from which things take their rise they pass away once more, as is meet; for they make reparation and satisfaction to one another for their injustice according to the ordering of time.
>
> Anaximander[1]

I

The greatest thinkers, from Anaximander onwards, have known that time is not merely a diverting topic, but a pervasive and hydra-headed problem. Indeed, much philosophy reads like the construction of seawalls against it. For time is the destroyer of all that we are proud of, including pride itself, and even threatens the realization of philosophy's highest ideals.[2] Time is the possibility of corruption at the deepest level. And yet without time's synthetic powers, without organized temporal extension, there would be nothing to be corrupted. Time makes as well as breaks. Time giveth and it taketh away.

But the pervasiveness of time in philosophy does not arise simply at the level of intuition – that all things are finite, that all experiences are in time. Its shadowy form lurks beneath the surface of most of the central problems of philosophy, and has a major impact on how we think of identity, of truth, of meaning, of reason, of freedom, of language, of existence, of the self ... the list could be extended indefinitely. And even talking of a major impact on these problems is too weak a verdict. It would be wrong to imagine these being happy, respectable, problems only subsequently visited by the scourge of time. Without time, these

Time-shelters: an essay in the poetics of time   225

would not be the problems they are at all. How a person stays the same person over time may be a problem. But would we be persons at all without the challenge this problem presents? In brief, time is not just a special problem, not just the proper object of research on the philosophy of time. The ubiquity and pervasiveness of time does not, however, as tidy minds might hope, make it discountable. If everything, in addition to its natural color, were tinted by a peculiar shade of pink, we *could* discount the fact, because we could neither detect the color, nor understand the meaning of this condition. But time is not the name of a simple predicate or homogenous condition. Nor is it simply the name of a neutral dimension in which independently identifiable things can be ranged, like birds on a telephone wire.

Is there then any overarching way in which we can start to think time? If we followed Anaximander, we could say that time is the *economy* of being.[3]

Anaximander's fragment has stimulated many different translations and readings, and I do not intend to enter into these matters here.[4] But it is clear that there are a number of different *levels* at which it could be read: (1) It is a vision of cosmic pathos: everything comes and goes. (2) It allows us to focus on the very phenomena of emergence into presence, lingering persistence, and withdrawal. (3) It points to a radical *heterogeneity* in time. Time involves not just a sequence of identical now-points, but also a plurality of *beings* whose being is dramatized temporally.

This latter perspective is the one I wish to pursue here. And I can begin to explain the title of this paper by recalling the title of an eccentric and delightful book by Gaston Bachelard. *The Poetics of Space* is a study of intimate spaces.[5] Bachelard combined, in the most remarkable way, a scientific and a poetic mind, and this book is dedicated to the description of a whole range of worldly enclosures – shells, cupboards, nests, houses, boxes, etc. It makes a powerful contribution to the project of providing alternative models to that of a single, neutral, geometrical Space, a project being pursued in so many ways – from Merleau-Ponty's phenomenology of embodiment to the recent renewal of interest in architecture.[6] But what would a poetics of time look like? And what use would it be?

The Poetics of Time could title a work celebrating the literary, existential and artistic variety of temporal forms. I will not be doing this directly here, but what I propose will bring such a work closer. Allow me to put forward two principles, each with powerful consequences for our philosophical practice:

1. That critique, deconstruction, revisionary metaphysics (however different from one another), each presuppose a prior level of philosophical engagement with phenomena that we will here call *interpretation*. They are not *substitutes* for it, though they may demonstrate its limits, or set up a backwash of new considerations for further interpretation, but they *do* presuppose it.
2. Such interpretation itself presupposes typically unthematised *schemata* of space and time. And these schemas play a crucial role in determining even the limited adequacy of our interpretations.

We are concerned here with pursuing a particular reading of the Anaximander fragment – in which time appears plurally in beings whose being is dramatized temporally. We will be providing a framework for the way in which time enters into the constitution of beings, not just Being – of things, events, complexes of relationships, institutions, persons and so on. With a glance back at Bachelard, in this poetics of time, I will call all such beings *time-shelters*. And, to anticipate the obvious question, yes, I accept that the distinction between space and time is, if anything, even more fragile than the boundaries that mark out beings from one another. Consider for example, the ear, its labyrinthine complexity dedicated to the requirements of sound and its transmission, or those shapes for charming time that we call musical instruments.

Time-shelters could be described as local economies of time. But why this term economy? It responds to two exigencies: (i) to preserve a metaphysical neutrality of the sort that Husserl's phenomenology attempted to sustain – in particular to belong neither to the pole of the object nor the subject; and (ii) to capture the possibility of formally describing the modes of constitutive and regulative management of the boundaries of things of all sorts, by which their being and identity is created and preserved. My sense is that the language of economy is as suitable a discourse as any in which to pursue ontological neutrality. Where what we call natural phenomena are being discussed, the claim is being made that they exhibit general properties which transcend their natural domain.

What is a time-shelter? Let me offer an account in a somewhat queer key: the universe as a whole is entropic. Human beings however – indeed all living beings – are essentially negentropic. In the midst of growing disorganization, we find creatures that consume and manufacture organization and complexity, whether at the molecular level or that of information. These frames generate internal boundaries within the world, which both establish and mediate the relationship of inner and outer. No model of the world as an homogeneous unity survives a

Time-shelters: an essay in the poetics of time

moment's scrutiny. And the presumption of a higher order synthesis of these lower level orders (as, for example, in Leibniz's *Monadology*) will here be suspended for want of plausibility. The implications of this for our thinking about time are profound. If there are semi-autonomous local economies which relate in each case to various 'outsides', then the boundary between inside and outside will function in much the same way as what has been described as an *event-horizon* at the 'rim' of the astronomer's black-hole. A boundary is not a thing, but a cluster of procedures for the management of otherness. Crossing the boundary brings about a discontinuity of both spatial and temporal relationality. Such discontinuities provide what we could call shelters for the growth of more luxuriant forms of spatial and temporal organization.

This phenomenon of sheltering has enormously wide ramifications but I would like to clarify its significance before offering some illustrations. Shelters are persistent forms of event-discontinuity in the world. They manage the boundaries of inside and outside by representing the outside within, by translating, buffering and anticipating that which impinges, by resistance, accommodation, expansion and exchange, and by establishing for these purposes rhythms, rules and regularities. But they are not, and cannot be, sealed from the outside, nor are they immune to total transformation or destruction. The boundaries of shelters are essentially permeable, in ways that allow interruption – invasion, infection, corruption. It may always be possible – I take no view on this – to give a location in world time to sheltered events. What is distinctive about shelters and their event-horizons is that such external correlation is often as tortuous as translation between languages.[7]

The implication of all this is not that there are (as there may be with black-holes) absolute boundaries sheltering local temporalities, but that such boundaries, at which are negotiated and managed the literal transactions across them, and which pose difficulties even for the virtual penetrations of description and calculation, are nonetheless real. The language of flux is not enough. Or at least, there are orders of flux.

Husserl's account of the phenomenology of internal time consciousness begins, innocently enough we might think, by a methodological bracketing-out of the world, and objective time, so as to focus on the indisputable temporality of inner time-consciousness – which would seem to survive even the destruction of all we know – as long as we can still dream, remember or imagine.[8] My initial appeal is no less intuitive, and indeed arguably satisfies Husserl's original requirement of ontological neutrality even more effectively. Most of the 'things' we see around us are shelters of the sort I have described – from football matches to musical performances, from chemical reactions to trees, from States to universities. What they have in common is that they each exhibit semi-autonomous temporal organization – not just organization in time, but of time.

But how are we to think of temporal structure and organization? Time is distinct from any appearance. It is not in itself anything we can have in front of us, present to us. When we do think about it, it is easy to identify it with the passage of time. We can witness the flowing of the stream, the rising of the sun, the movement of the hand on the clock. If we think of time simply as change, then clearly we have some direct experience of it. And some such object or content of consciousness is essential. Sheer empty consciousness, were it imaginable, would be an eternal present. Temporal structure is however available to us more indirectly, through music, for example, which provides us with powerful intuitive access to the idea of temporal structure. Music, we could say, is the dramatization of time in all its variety, and that very variety – from fugue to Cage – gives us some intuitive access to times and temporal structures in the plural. The awareness we have of our own bodily rhythms' pulse, breathing, sleeping and waking, hunger and satiation, undoubtedly functions as something like a clock, the metronome that each of us is, a primitive somatic synthesizing of time. It is surely such rhythmic interlacings that a phenomenological theory of temporal constitution takes for granted, so that we always arise up and find ourselves not just as heavy bags of skin set off against the world, but in rhythm. If music is, in this sense, a derived experience of time, dependent on the temporal organization of our embodiment, it allows us for the first time a glimpse of these structures, precisely through the operation of aesthetic distance. I hear the music – I am not identical with it as I am with my own rhythmic jungle.

For Plato, art directs us to timeless forms, but music directs us to the forms of time. For all the importance of tone, timbre, and other intrinsic qualities of sound, music also exemplifies a radical reduction of the real – sensory reduction, selectivity, separation from objects, even persons. Music is perhaps the best intuitive source we have of temporal structure. It exhibits variation, development, co-ordination, interference, interweaving, recapitulation, as well as all the psychological dimensions associated with memory and expectation. It combines the operation of rules, rhythms and repetition to occupy in an exemplary way that middle ground in which subject and object are intimately entangled. And if this is so, we find in the structure of music the same phenomenon of double mapping that characterizes the traditional concept of the person, as a being subject both to rules and rhythms, principles and passions. Although I shall not pursue the matter here, it is clear that the interference patterns between these two types of temporal structure generate many of the problematic self-understandings we find in traditional reflections on the self. More generally, openness to the entanglement[9] of distinct and conflicting times, plots, even logics is a powerful hermeneutic principle when thinking about people, texts, history and even natural phenomena (e.g., disease, atmospheric changes, economic cycles).[10]

It is important to clarify the status of this account of time-shelters. It operates primarily at what Heidegger would call the ontic level, rather than the ontological. And I believe it conforms to powerful intuitions about what entities in general are – relatively autonomous local economies of time (and, it has to be said, space).[11] Identity, on this model, has a great deal to do with such relative autonomy, such semi-permeability to what is external, and is something which has to be sustained and developed rather than being given. I take such an account not merely to be a description of the way things are, but to have hermeneutic power, the results of which will confirm its value as a model.

But I did not begin with Anaximander for nothing, and in this account of time–shelters, there is clearly an adumbration of the ontological as well, for it offers a very general account of what it is for anything at all to be. As we will see later, this will generate its own 'beyond being'. Yet to talk about the ontological is not necessarily to talk about totality. What does this account of time-shelters do for the traditional problem of whether time is one or many? To pursue the project of articulating local temporal economies, we have to suspend the thesis of the unity of time, to put it out of action, as Husserl would say, just as one has to suspend one's ordinary visual relation to the world when looking through the microscope, and one's ambient auditory awareness when listening to music. We suspend the thesis of the unity of time in the following sense: we no longer take for granted that all times will line up in a single series, that all series have the same form, and even that all time is serial. It may be argued that this thesis nonetheless has a regulative value. I think it does, but the recognition of such a regulative value is squandered if it remains an *a priori* reassurance.[12] It is far more productive to interpret the unity of time as a principle of boundary permeability, which could be expressed in the following way:

All beings are realms in which a certain more or less powerful order holds sway. Each order constitutes a boundary, hence an inside and an outside, or a laminated set, or entangled matrix, of such insides and outsides. These boundaries are not, however, primarily spatial, but economic; and insofar as economy is concerned not just with permanence and continuity, but also destiny, transformation, dehiscence and conflagration, then what is at stake is always time.

If, as I am claiming, local economies generate discrete temporal interiorities, then the thesis of the unity of time would change its modality. It would become a claim about the permeability, corruptibility, and fragility of any autonomous time. Not all that is fragile breaks, and the vulnerability of all boundaries to breaching does not license temporal totalization.[13] The pianist always may collapse in the middle of a live performance, but most often, human mortality allows the concert its own time.[14]

II

All this is offered as a contribution to a hermeneutics, not just an ontology. There is, however, a great gap in this whole account, a massive lacuna. I have claimed that beings of all sorts are time-shelters, that they exhibit, contain, and are constituted by local temporal orders, between which there can be both allergy and respect. So far we have given little idea of how these responses arise, how they are possible.

I have said that I prefer the language of economy for its ontological *neutrality*, which licensed our putting to one side the question of the origin of time-shelters or the extent of our contribution to their emergence, maintenance and disappearance. That we could successfully even begin without considering such matters does, of course, betray a certain concealed ontological commitment – that these local economies that we call beings have a certain autonomy in relation to individual human beings.

Let me be explicit: I am not considering the universe as it might somehow be in-itself, its noumenal reality. Nor am I considering it in the idiosyncratic light in which it appears to this or that subject. Our account of time-shelters can, for the most part, be understood as an account of a dimension of our thrownness,[15] what we find ourselves already in, confronted with, surrounded by, ways in which the real is already constituted, determined, articulated, ways we take over and adopt as our own, and in so doing come to be subjects in the first place. Time-shelters are not just *subjective* phenomena, but describe the umbrella form of what *we* take to be real.

Time-shelters: an essay in the poetics of time 231

What is it then that is so resoundingly mute in our account? The name we commonly give to that field of meaningful structures in which we find ourselves, is language. To draw language successfully into the temporal constitution of beings is something I cannot begin to tackle here in any general way. Even in individual words, language names, honours, creates, protects, holds open and preserves the beings we have described as sheltering and being sheltered by local time. Through *poetry,* language can bring these realms to presence, can make manifest the fragility of their lingering between coming into being and passing away. And in what Aristotle called his *Poetics,* it is through those dramatic constructions of language that we call tragedy that our own finitude is taken hold of and worked through. Language as *poesis,* as making and constituting, is inseparable from the appearance of such time-shelters.

There is, however, a particular way of pursuing this connection between time and language that I would like to discuss briefly here – one in which it is the being of being-human itself which is in question – that remarkable intellectual adventure with which Paul Ricoeur presented us in *Time and Narrative.*[16]

The site of Augustine's perplexity about time was that of reflection. And in view of the subsequent efforts of philosophers, one might think that there was something about reflection that doomed our thinking about time not just to initial confusion but to final failure. In this book Ricoeur takes up the challenge and singles out as the persistent difficulty in the history of philosophy that of reconciling, of doing equal justice, to both phenomenological and cosmological time. In the story he tells, Aristotle's understanding of time as 'the number of movement with respect of the "before" and "after"'[17] gives us no way of thinking about time as experienced, even if an apprehending subject is actually required on Aristotle's model.[18] Augustine's account of the distension of the soul,[19] while it addresses that problem, does not offer us any basis on which to think of objective time. And he needs to be able to account for this, because he wants to say that time began at the point at which the world was created. Kant, Ricoeur suggests, may give to the mind the role of being the condition of worldly time, but gives us no phenomenology of experience. While Husserl, for his part, has great difficulty in reconstituting objective time after the *epoché.*[20] Finally Heidegger's attempt to think of world-time as a levelled-off authentic time is a failure, because even this world-time, Ricoeur claims, is specific to individuals.

I will not, on this occasion, comment on the details of this story. What it indicates is the scope of Ricoeur's concern about time. His suggestion seems to be not merely that time is pervasive, as I have claimed, but that Augustine's perplexity about time reflects a deeper failure to think time: not just Augustine's failure, but the failure of a tradition.

Time and Narrative is structured around a hypothesis – which in mock-Heideggerean idiom Ricoeur sums up at one point as the claim that narrative is the guardian of time. More carefully expressed, it is the claim that the phenomenology of time, found, for example, in Husserl's *On the Phenomenology of the Consciousness of Internal-Time*, 'the most exemplary attempt to express the lived experience of time', leads to the multiplication of aporias. And these tangles only get unravelled through 'the mediation of the indirect discourse of narrative.' One can hear Kant's claim that intuitions without concepts are blind in Ricoeur's insistence that phenomenology must embrace the conceptual resources of language. But he is going further and saying that what Kant called schematism, in his 'Analytic of Principles', the application of a concept to an instance by a productive rule-governed imagination, need not be thought of as 'an art concealed in the depths of the human soul whose real modes of activity nature is hardly likely ever to allow us to discover and to have open to our gaze.'[21] It can be found 'writ large' in narrative.[22] Through narrative, successive events are subjected to *configuration,* a concept that generalizes Aristotle's account of plot in his *Poetics*. Narrative heals aporia.

But Ricoeur's extraordinary power and virtues as a thinker emerge most strongly in the *Conclusions* chapter to *Time and Narrative III*. Here he reviews the argument of the book, and as if the ravages of time and reviews could not be relied on to do it for him, he begins to unravel his own knitting. The hypothesis that narrative relieves us of the aporetics of time is seen to have limits. One of the central products of narrative is to allow us to construct a narrative identity – both at the level of history (and e.g. the identity of a nation),[23] and at the level of the individual life. This represents a considerable advance over accounts based on substance, or bodily continuity, or memory. Ricoeur had approached this idea earlier in his book on Freud, in which psychoanalytical treatment, what Freud himself had called 'the talking cure', is seen to culminate not in psychic readjustments but in the construction of a story that makes sense of one's own life, or especially those parts of one's own life that would otherwise be unthinkable.[24] Narrative draws in and allows mediation between causal and intentional time, it encompasses both events that happen to me, and actions I perform. In this way it deploys our most powerful synthesizing abilities in constituting an identity. And insofar as

Time-shelters: an essay in the poetics of time

it deals with both events and actions, it could be said to transcend the aporetic opposition between cosmological and phenomenological time.

In our idiom, Ricoeur's account of narrative when applied to the constitution of personal identity, manages the boundary of the self – excluding what will not fit, articulating what will. A person is not just a body; but surely the need to maintain boundaries is presupposed by any concept of personal identity – the idea of an individual substance would be only a magical solution.

But this hypothesis is nevertheless open to the objection that it makes identity somewhat unstable, insofar as many stories can be woven from the same material. Ricoeur treats this not as an objection but as a limitation – a distinction to which I shall return. We might equally regard it as an advantage to have a model which can accommodate the contingency and revisability of identity, a model which is not an all-or-nothing solution. Ricoeur also admits another, perhaps more worrying problem. Narrative identity *stresses the intelligible organization of events* at the expense of the will, the ethical moment, the moment of decision, of impetus.[25] In each of these examples, narrative does not just heal, it opens new rifts – first, the irresolvable plurality of stories, and then second, the opposition between the organizing power of imagination, on the one hand, and the will on the other. Ricoeur's account of the significance of these difficulties is classic. Narrative does not *resolve* aporias, it makes them *productive,* which suggests that a formal or logical solution to our problems (e.g. McTaggart's proof of the dependence of the B-series [cosmic time] on the A-series [lived time][26]) may not be required even if it were possible. His further, Kantian, gloss on these difficulties is to say that the limits of his account of narrative can be seen not as a defect but as 'circumscribing its domain of validity.'[27] But another reading of the appearance of these limits is possible: that they represent the return of the repressed, the re-emergence of the aporetic dimension that narrative was hired to keep under control. Limits of validity mean: beyond this point, unintelligibility, contradiction, aporia. Is not Ricoeur putting a brave face on time's reassertion of its power to disrupt all attempts at conceptual domestication?

As if this were not bad enough, Ricoeur discovers that the aporia generated by the gulf between cosmic and phenomenological time, a gulf bridged by narrative, the aporia that occupies centre stage in *Time and Narrative*, is in fact only one of many aporias to which our thinking about time is subject. The second is precisely that of time as one and as many, not in the sense of local time-frames that we discussed earlier, but rather in the sense that we think of time as one and yet as divided between past, present and future. Without rehearsing the argument here – he makes a subtle deployment of the idea of the unity of history – suffice it to say that Ricoeur finds narrative even less able to deal with this difficulty. The third and last aporia he designates 'the inscrutability of time', by which he refers to the various ways in which time continually breaks through our attempts to constitute it, to clarify its meaning, to show us its deep archaic enveloping mystery.

> What fails is not thinking, in any acceptation of this term, but the impulse – or to put it a better way, the *hubris* – that impels our thinking to posit itself as the master of meaning ... [T]ime, escaping our will to mastery, surges forth on the side of what ... is the true master of meaning.[28]

Now in my judgement, the true master here is Ricoeur himself who, after the fact, and after Augustine, has recast the plot of *Time and Narrative* as a confession, in which the presumption of synthesizing thought is confronted by a power that exceeds it. If time ever seems to be acquiescing in our configuring plots, you can be sure it is silently gathering its forces for revenge. Of course the oddest paradox now arises. For we have made time into the hero of a story of confinement and release. Has not narrative finally closed the trap, and triumphed?

In the account we have given so far, we have sketched the path of Ricoeur's own reflections on his attempt both to bind and to illuminate time through narrative. But the third aporia he discusses – the inscrutability of time, its power to reassert its envelopment – surely opens another front: the wider relation between time and language. One response to the breakdown, or the coming up against limits, of the power of narrative to tame time, might be to reassess the specific theory of language to which Ricoeur is committed in discussing narrative. To put it very simply: might it not be that narrative is committed to the possibility of a *certain closure of meaning,* which will inexorably be breached. In other words, narrative selects from, but does not exhaust, the power of language to resolve the aporias of time. Its particular forte is synthesis. But it has no monopoly on linguistic synthesis, and, more particularly, it may be just such a strength that is its weakness. If time is not essentially captured by the effects of closure that narrative facilitates, then we must either find time showing through in the very pathos of narrative's failure, or we must look elsewhere. We might expect that resources for such an expansion could be found in metaphor and metonymy, which seem to allow us to think the non-linear, creative, interruption of that articulation

Time-shelters: an essay in the poetics of time 235

of sense through time that we call narrative. We recall that the study of metaphor and narrative are for Ricoeur integral parts of a general poetics, 'one vast poetic sphere', and both instances of the productive imagination. We have already seen him describe narrative in terms of production; we know that Ricoeur, following Aristotle who thought of plot as the mimesis of an action, allows the poetic a role in the narrative refiguring of action; we know that Ricoeur ultimately seeks to harness the poetic for speculative and eventually practical ends. Without undertaking here a full-scale review of *The Rule of Metaphor*, it would not misrepresent Ricoeur to conclude that his deployment of metaphor and narrative separately and in harness is subordinated to a law of productivity, in which the moment of synthesis has the last word. If this is right, then we will not be able to find in metaphor a countervailing force that would interrupt narrative, or set up different trails of connectedness.[29]

III

My discussion of narrative has been framed by the problem of the aporias of time. Ricoeur introduced narrative poetics as an antidote to the multiplication of aporias in a pure phenomenology of time divested of these resources of linguistic synthesis. But as if by a process of selective mutation, new aporias continued to erupt. Should we not treat Ricoeur's analysis, and his extraordinary confession, as evidence of time's recursive power of interruption of our best-laid plots and plans? Ricoeur, you will recall, hoped that narrative time would heal the aporias of time, in particular its diremption into phenomenological and cosmological time, in which the subject/object opposition flourishes unhindered.

On another occasion, I will develop further some of the material in *The Deconstruction of Time* in which I argue from a somewhat different direction for a convergence of these two limbs of time. I claim that the radical difficulty of thinking phenomenological and cosmic time together is a consequence of the thesis of the unity and linearity of time. This is an a priori assumption that neither the natural nor the interpretive sciences bear out. I claim that the complexity of the temporal structures that arise from the ashes of the thesis of unity takes us a long way towards convergence. The first part, the current essay, dealing with time-shelters, exemplifies such a convergence. And the second part, focusing on Ricoeur's account of *Time and Narrative,* confirms such an approach, even as it runs up against limits. I too would now like to turn to the question of limits, to take up again the way Ricoeur *presents the issue*, and to attempt a somewhat different formulation.

The thesis of the inevitable return of aporia, which Ricoeur does not formulate, but which is not too difficult to erect on the evidence of his work, might be thought to have a simple explanation. If references to phenomenological and cosmic time are not just to partial models of the real, but to discrete and autonomous dimensions of the real, then any being subject to double mapping, subject to constitution by both of these forms of time, would face not just an intellectual difficulty in reconciling two partial descriptions of the same thing – like the Evening Star and the Morning Star – such a being would suffer diremption, and the wound could never finally be healed. Such a being, it might be said, is man. The quest for reconciliation following, most notably, Kant, typically involves the subordination of causality (cosmological time) to self-legislation (phenomenological time), or at least the establishment of the independence of the latter from the former. If I am right, however, this account of diremption rests on an opposition which the setting aside of the unity of time thesis would dramatically weaken. My account, however, makes things worse, not better.

I have tried to suggest that we think of time plurally, and structurally, in various complex ways. But this entire approach, even if valid, seems to me to be limited. The accounts we have given of the economy of local time-shelters, and Ricoeur's accounts, as we have relayed them, of narrative and of narrative identity, are analytical, neutral – I claimed that this was an advantage – and involve a certain 'distance'.

The stage for the re-emergence of this distance, which echoes the split between causal and phenomenological time, can be found, amazingly enough, in book VIII of Aristotle's *Poetics*. There, Aristotle writes:

> Unity of plot does not … consist in the unity of the hero. For infinitely various are the incidents in one man's life, which cannot be reduced to unity … That is the error of all poets who imagine … that as Heracles was one man the story of Heracles must also be a unity. Homer … seems happily to have discerned the truth … In composing the *Odyssey* … and likewise the *Iliad* [he made them] centre around an *action* that … is one. [my emphasis][30]

The construction of a narrative is an aesthetic unity, whether of a human life, or of a nation. If Aristotle is right, then even the supple connective resources of narrativity may be doomed to a certain failure because there will always be bits that belong but do not fit and never will. We may even conclude that the narrative quest for identity is a dangerous one, exhibiting at a more plausible level the same rigidities of self-understanding as we left behind when we left behind the idea of substance.[31] But we might equally conclude differently that the demand for some sort of temporal intelligibility is unavoidable for humans, even as we recognize it to be unfulfillable. Earlier, when describing the order that holds sway in all beings, I said that each order constitutes a boundary, hence an inside and an outside, or a laminated set, or entangled matrix, of such insides and outsides. Our survival, day-to-day, does not usually depend on narrative intelligibility being successfully maintained. What we call our embodiment, the maintenance of our corporeal integrity, not to mention our struggles for truth, or for minimal control over the forces that surround us – all these are conjoined and entangled with narrative in an economy in which a multiplicity of times and rhythms intermingle.

It would be my hope that such a description, for all its complexity, would actually make interpretation, if not easier, at least less liable to over-simplifying. But might it not be that, again, this whole account can be circumscribed within a fatal limitation?

In the account we have been offering, time has, on the whole, been presented positively, constructively, economically, and some such account as this seems indispensable. But what does it tell us about what it is *to be* a being described in this way – one that manages the maintenance of fragile and permeable boundaries? Can we reduce madness and mortality to questions of economy? They are surely not theories, but nor are they simple conditions we must endure. The language of economy is no attempt at their theoretical evaporation; however, this language might be indispensable for us to get any handle at all on such mortal issues. Ricoeur's last aporia centred on the inscrutability of time. We should perhaps remember here that time is not just an enveloping beyond to all our little bubbles of order. For all our ability to breed domestic forms of time, it also holds in reserve the apocalyptic possibility of dissolving any and all of the horizons of significance we have created for ourselves.[32]

What, then, after all this, of the ethical? If Anaximander spoke of 'justice', surely, as Theophrastus first complained, this is misplaced poetic metaphor – what we have here is a cosmology, a cool, detached ontology. And do we not fall foul of Levinas' critique of the neutrality of ontology?

I would say this: a discourse does not become ethical by simply speaking of the other, but in the ways it offers itself to appropriation. It is precisely the understanding of the fragility and vulnerability of all beings that both opens the space of ethics and prevents its premature humanistic closure. The best appropriation of this text would be one that enhanced respect for the miracle of diverse beings, including oneself, and for the variety of forms of survival and expenditure. At this point, poetics would become ethics.

Department of Philosophy
Vanderbilt University
Nashville, TN 37240, USA

Notes

1 According to Theophrastus, preserved by Simplicius. Translation cited from John Burnet, *Early Greek Philosophy* (London Macmillan, 1892; & New York, Meridian Books, 1960), p. 52.
2 Most obviously, for Plato, truth. I have a Nietzschean perspective in mind here.
3 Why use the word economy? Time is usually understood formally or abstractly in terms of seriality, or 'falling-off' [*Ablaufsphänomene*]. But if we put aside this abstract description, what open up are not just the more archaic concerns of the pre-Socratics, but also further formal properties of time, as it functions in identity for example. The language of economy serves to bring about such a displacement, and allows us to *harness* time.
4 At this and at other places in the essay, there are obvious Heideggerean filiations and resonances. If I do not here discuss his essay 'The Anaximander Fragment', [in Heidegger, *Early Greek Thinking*, trans. D.F. Krell & F.A. Capuzzi (San Francisco: Harper & Row, 1984); '*Der Spruch des Anaximander*' (1946), collected in *Holzwege* (Frankfurt am Main: Vittorio Klostermann, 1963)] it is not for want of interest. Rather I am trying here to develop a more analytical conceptual matrix with which to articulate some of the same issues, rather than succumbing to the pleasures, pains and inevitable seduction of reading Heidegger.
5 Gaston Bachelard, *The Poetics of Space*, trans. M. Jolas (Boston: Beacon Press, 1969); *La poétique de l'espace* (Paris: Presses Universitaires de France, 1958).
6 Of special note is Ed Casey's *The Fate of Place: A Philosophical History* (California: University of California Press, 1999).
7 For the following kinds of consideration:
   (i) There may be no recognized calibration of the iternal event. (When exactly does the bud open? We may be able to say after t_1, before t_2.)
   (ii) The event itself may depend for its distinct identification on its internal relations to other events within the same shelter. Trans-boundary temporal location may generate incoherence, or at least oddity. (Christopher Columbus discovered America at 3:32 in the afternoon).

Time-shelters: an essay in the poetics of time 239

(iii) The value and significance of external and temporal measurement may be extremely limited compared to the internal temporal order. (A mother's argument against chemically induced labour to suit the hospital timetable.)

8   Edmund Husserl, *On the Phenomenology of the Consciousness of Internal-Time (1893–1917)*, trans. J.B. Brough (Dordrecht: Kluwer, 1991); *Zur Phänomenologie des inneren Zeitbewusstseins (1893–1917)*, ed. R. Boehm (*Husserliana X*; The Hague: Martinus Nijhoff, 1966).

9   Cf. Merleau-Ponty's note criticizing Husserl's tendency to seek to '"disentangle", "unravel" what is entangled', and his subsequent claim that the 'idea of chiasm and *Ineinander* is on the contrary the idea that every analysis that *disentangles* renders unintelligible'. *The Visible and the Invisible*, ed. C. Lefort, trans. A. Lingis (Evanston: Northwestern University Press, 1968), p. 268; *Le Visible et l'invisible* (Paris: Editions Gallimard, 1964), p. 321.

10 Freud's discussion of the reality principle, the pleasure principle, and the drive towards oblivion – Thanatos – would be an obvious case. And it is revealing that Freud is one of the thinkers who have explicitly used the language of economy (see, for instance, in the collection *On Metapsychology* [vol. 11 of the Penguin Freud Library, trans. J. Strachey (London: Penguin, 1984)], the essays 'The Unconscious' (p. 184), *Beyond the Pleasure Principle* (p. 275), and *The Ego and the Id* (p. 352)).

11 Why, it might be asked, do we persist in talking about time, rather than time and space, or time-space? First, it is harder to introduce time into spatial thinking than vice versa. Indeed, the very linguistic articulation of time begins this latter process (see, for example, Bergson, *Time and Free Will*, trans. F.L. Pogson (London: Macmillan, 1910), pp. 131ff; *Œuvres*, ed. A. Robinet (Paris: Presses Universitaires de France, 1959), pp. 87ff). It may well be that all thinking about time ends up (as Heidegger does) with something like time-space, or some other primordial matrix that sustains both space and time. But I am convinced that the long haul of discovering that this is so is preferable to the short-circuit that knows it in advance. This account of time-shelters can indeed be read as a partial explication of why this path to time-space may be unavoidable.

With Heidegger's discussion of time as 'the *horizon* of being' in *Being and Time* (trans. J. Macquarrie & E. Robinson [Oxford: Basil Blackwell, 1962], pp. 19, 38, and *passim*; *Sein und Zeit* (Tübingen: Max Niemeyer Verlag, 1953), pp. 1, 17, and *passim*), there begins a long meditation on the intimate mutual imbrication of time and space, culminating, perhaps, in 'Time and Being' (1962), and punctuated by sustained discussions on Time-Space in the *Beiträge zur Philosophie* (written 1936–8), §§ 238–242 (translated by P. Emad & K. Maly as *Contributions to Philosophy (From Enowning)*, [Bloomington: Indiana University Press, 1999]). It was only after completing this paper that I came across Heidegger's discussion of Truth and Sheltering

(§§ 243–247), in which he makes a not unrelated use of the idea of sheltering, though one tied into the problematic of truth as concealment. Exploring this connection is work for another day.

12 For an extended treatment of such questions, see David Wood, *The Deconstruction of Time*, (reprinted with new introduction; Evanston: Northwestern University Prss, 2001).
13 See G. Deleuze and F. Guattari, *Anti-Oedipus: Capitalism and Schizophrenia*, trans. R. Hurley, M. Seem & H.R. Lane (New York: Viking Press, 1977); *L'Anti-Oedipe* (Paris: Les Editions de Minuit, 1972). They claim that capitalism deterritorializes and reterritorializes every domain or place or unity enclosed by a threshold or boundary. What I claim for time-shelters is that they are sites of resistance to such transformation. And if every time-shelter is *vulnerable* to such invasion, nothing requires that an incursion will *always* take place.
14 I recall, for example, a news report of a man who was allowed to finish his breakfast before being formally arrested and driven off to jail.
15 See Heidegger, *Being and Time*, § 29, and *passim*.
16 Paul Ricoeur, *Time and Narrative* (3 vols.), trans K. Blamey & D. Pellauer (Chicago: University of Chicago Press, 1988); *Temps et Récit* (Paris: Editions du Seuil, 1985).
17 Aristotle, *Physics* IV, 219b2.
18 Aristotle, *Physics* IV, 218b21ff.
19 Augustine, *Confessions*, trans R.S. Pine-Coffin (London: Penguin, 1961), chap. 11, especially §§ 14ff.
20 The point of departure for Husserl's phenomenology of time is the *epoché*, or bracketing, of objective time. Husserl's aim is to focus his enquiry on the way in which time is given as a phenomenon of consciousness.
21 Kant, *Critique of Pure Reason*, trans. N. Kemp Smith (London: Macmillan, 1929), A141/B180–1.
22 For an extended treatment of this Kantian connection, see Kevin Vanhoozer, 'Philosophical Antecedents to Ricoeur's *Time and Narrative*', in David Wood (ed.), *On Paul Ricoeur: Narrative and Interpretation* (London: Routledge, 1991).
23 See J. Habermas, 'Concerning the Public Use of History' in *New German Critique*, vol. 44 (Spr/Sum 1988), pp. 40–50.
24 Paul Ricoeur, *Freud and Philosophy: An Essay on Interpretation* (New Haven: Yale, 1970).
25 Ricoeur alludes to Levinas on promising, but he could equally have cited Heidegger.
26 J.M.E. McTaggart, *The Nature of Existence*, vol. II (Cambridge: Cambridge University Press, 1927), Book V, Chapter 33.
27 Ricoeur, *Time and Narrative*, vol. 3, p. 261.
28 *Ibid*.

Time-shelters: an essay in the poetics of time 241

29 Resources for such interruption can be found in the work of Derrida, Heidegger, Blanchot, Levinas and others. It is worth looking at an attempt at a synthesis of these resources in a recent book by Herman Rapaport – *Heidegger and Derrida: Reflections on Time and Language* (Lincoln: University of Nebraska Press, 1989). He shows that the trails of linguistic *association* set up in texts (and in life) by such tropes as paronomasia and metalepsis provide a kind of involuntary counterpoint to the harmonies and melodies of narrative organization.
30 Aristotle, *Poetics*, trans. S.H. Butcher, (New York: Hill and Wang, 1961).
31 Paul Ricoeur, *Oneself as Another*, trans. K. Blamey (Chicago: University of Chicago Press, 1992); *Soi-même comme un autre* (Paris: Editions du Seuil, 1990). His central claim is that selfhood, or personal identity, does not require the kind of identity we associate with an unchanging self-sameness. Moreover, such a self is tied to a productive dialectical engagement with otherness and the other. Ricoeur's account gives a stunning display of how what we have called the 'management of boundaries' might be thought to operate through language, action, and reflection, and the implications of this for ethics and ontology.
32 I allude here to Husserl's discussion of the possibility of the end of the world, the destruction of all sense (in *Ideas Pertaining to a Pure Phenomenology and to a Phenomenological Philosophy*, First Book, trans. F. Kersten [Dordrecht: Kluwer, 1982]; *Ideen zu einer reinen Phänomenologie und phänomenologischen Philosophie, Erstes Buch* (*Husserliana* III, ed. W. Biemel; [Den Haag: Martinus Nijhoff, 1950]; § 49), taken up so interestingly by Levinas in his 'Simulacra: The End of the World', in David Wood (ed.), *Writing the Future* (London: Routledge, 1990); '*Simulacres*', in *Effets de neuter*, vol. 1 (1985), pp 9ff.

Index

accident 45, 68, 71, 72, 77, 78, 158, 170, 193, 194, 202, 203, 213n8
act 31, 32, Ch3 *passim*, 67, 69, 70, 71, 79, 82, 87, 89, 90, 92, 151, 156, 169, 193–198, 200, 204–206, 233, 235, 237, 241n31
actuality of 51, 70, 89
Adamson, G. 157, 164
adjacent possible 137, 138
Anaximander 3, 224, 225, 229, 237
Ansell Pearson, K. 20, 21, 24n15
Aristotle 1–19, 21n2, 22n8, 23n11, 23n13, 24n15, 144, 190, 191, 193, 205, 215n34, 231, 232, 235, 236
  magnitude in 12–16
  stretch of time in 15, 17, 18
association 25, 43, 57, 103, 194
Atomism 20, 80, 191, 193, 204–211
atoms of time 5, 6, 12, 13
attention 49, 57, 58, 60, 69, 78, 79, 100, 105, 193–197
  to life 42, 45, 58
  inattention to life 60, 61
Augustine 231, 234
awakening 43, 46, 86, 202, 203, 205, 206, 215n27

Bachelard, G. 2, 19–21, 96, 102, 144–147, 149, 168, 170, Ch9 *passim*, 225
  The Dialectic of Duration 92n1, 94n6, 94n10, 95n16, 192, 197, 198, 211
  L'Intuition de l'instant 92n1, 93n2, 95n17, 192–198
  Poetics of Space 21, 225

horizontal and vertical time in 196–198, 213n13
Bacon, F. 80
Badiou, A. 144, 145, 168, 176n29
Barbour, J. 2, 3, 19, 21, 24n23, 112, 113, 120, 135, 140n11
  The End of Time 19, 96, 102, 105, 106, 112, 119, 125
becoming 7–12, 23n14, 24n15, 49, 67, 71, 73, 74, 77, 83, 99, 154, 164, 169, 172, 193, 201
beginning 67, 71, 72, 79, 199, 200
Bergson, H. 1, 2, 19–21, 24n15, Ch4 *passim*, Ch5 *passim*, Ch7 *passim*, 191–193, 239n11
  Time and Free Will 93n4, 94n7, 95n21, 146, 239n11
  Matter and Memory 47, 51, 63n36, 63n42–44, 63n52, 172
  Introduction to Metaphysics 172
  Creative Evolution 71, 145, 148, 149, 153–155, 158–160, 167, 168, 170, 171, 191, 215n24
  Mind-Energy 61m1, 94n4
  Duration and Simultaneity 96, 99, 104, 106–109
  Creative Mind 150
  tendencies in 20, Ch7 *passim*
biology 93n2, 116, 131, 137, 153, 158, 162, 163, 166, 170, 171, 207, 212n7, 219
Bohr, N. 95n20, 135
Boltzmann, L. 103–105, 184, 186, 219, 220
Borges, J.L. 162, 163
boundaries 117, 177, 180, 187, 226,

Index

227, 229, 230, 233, 237, 238n7, 240n13, 241n31
Butler, S. 64
Cage, J. 228
calculus 24n17, 91, 178
Carlyle, T. 87
Casey, E. 238n6
chance 91, 92, 120, 190, 194, 195, 198
change Ch1 *passim*, 24n16, 88, 91, 98–100, 103, 105, 133, 137, 138, 152, 155, 158, 169, 172, 177, 178, 180, 186, 190–192, 194, 202, 208, 209, 218, 219, 228
chaos 127, 218–222
clinamen 20, 208–210
completeness 3, 39, 42, 50, 54, 65, 69, 71, 77, 87, 106, 134, 148, 154, 158, 170, 171
complexity 20, 21, 32, 33, 38, 45, 46, 69, 70, 80, 103–105, 116, 125, 137, 145, 152, 164, 167, 170, 190, 191, 194, 196–199, 203, 206, 207, 211, 218, 226, 235
coincidence 40, 53, 76, 78, 79, 82, 100, 119, 126, 139n7, 194–198, 200–202, 209
configuration 101, 104, 117, 118
space Ch6 *passim*, 140n11
of universe 102, 117, 118, 131
consciousness 25, 26, 30, 32, Ch3 *passim*, Ch4 *passim*, 96, 103–105, 108, 109, 152, 154, 167, 170, 178, Ch9 *passim*, 227
springing-up of 48, 58, 68
contingency 149, 170, 171
continuous [*suneches*] Ch1 *passim*, 21n2, 23n13, 37, 40, 41, 46–48, 58, 66, 67, 70–72, 76, 80, 82–86, 92, 128, 133, Ch7 *passim*, 177, 188n1, Ch9 *passim*

and contact 4, 5, 10, 12, 21n2, 209
indivisibility of 4, 5, 13, 21n2, 146, 168
as undifferentiated 4
Copernicus, N. 97
Couturat, L. 83, 95n15
CP violation 184
Cramer, J.G. 20
creativity 2, 64, 68, Ch7 *passim*, 234

Darwin, C. Ch7 *passim*
Deleuze, G. 144, 145, 169, 173, 174n20, 176n29
Democritus 207, 208
Dennett, Daniel Ch7 *passim*
depersonalisation 36, 63n41
de-phasing 202–204, 210
Descartes, R. 79, 149
cogito 79, 197–199
deviation of atom 20, 190, 201–211
diffeomorphism 114, 115, 121, 123, 132, 133, 139n7
invariance 115, 119, 126, 128, 133
difference 17, 18, 48, 49, 100–102, 157, 203, 210
Dirac, P. 102, 181, 188n4
discontinuity 19, 21, 23n13, 25, Ch4 *passim*, 149, 159, 168, 170, 190–194, 205, 206, 212n6, 227
dissipative structures 219
dreams 36, 37, 39, 42–46, 53, 59, 61, 65, 68, 197
duration 219, 24n15, Ch4 *passim*, 98–103, 106, Ch7 *passim*, 174n7, 174n20, 192, 193, 195, 201, 206, 209, 212n5, 212n7, 212n12
heterogeneity of 74, 76, 129, 151, 170, 192, 212n6
homogeneity of 67, 77, 168
Durie, R. 24n22, 96, 212n6, 212n21

Einstein, A. 1, 2, 19, 74, 75, 95n20,

99, 101, 102, 106–109, 115, 116, 127, 178, 181, 186, 213n10
energy, conservation of 29, 159, 187, 218
entropy 104, 161, 184, 185, 218–220, 226
equilibrium states 103, 219, 220
ether 75
event 9, 10, 24n15, 34, 35, 43, 55, 63n43, 68, 70, 72, 76, 77, 80, 81, 83, 91, 102, 126, 128, 130, 135, 138, 144–157, 168, 176n29, 179, 182, 189, 190–195, 198, 205, 206, 209, 226, 227, 233, 238n7
event-horizon 227
evolution 121, 137, 139n8, Ch7 *passim*, 219
eye 57, 58, 96, 155, 159, 160, 170, 171

false recognition 2, Ch3 *passim*
Feynman, R. 182
field 114–116, 123, 127–129, 159, 179, 186
finality 148, 158, 171
finitude 2, 25, 166, 210, 222, 231
Finnegan's Wake 163
La Fontaine 86
freedom 53, 68, 73, 94n4, 98, 103, 188, 224
Freud, S. 232, 239n10
future 4, 7–9, 14, 16, 17, 25, Ch3 *passim*, 64–67, 76, 87–90, 103, 138, 145, 151, 156, 157, 177, 178, 180, 183, 184, 186–188, 195, 203, 218, 222, 234
  closedness of 60
  openness of 157, 222

Galileo, G. 144, 150, 218
Gell-Mann, M. 135

geometry 90, 92, 97, 102, 129, 138, 148, 171, 177, 179, 209, 211, 225
  Euclidean 97–99
  non-Euclidean 99, 102
Guyau, J.-M. 89, 94n9

habit 42, 52, 53, 60, 71, 89, 90, 93n2, 95n18, 148, 150, 154, 155, 161, 167, 172, 213n8, 214n16
Halbwach, M. 78, 94n10
halt in impetus of consciousness 59, 60
hamiltonian 122–128
Hartle, J. 135, 141n25
Hawking, S. 112, 184
Heidegger, M. 1, 18, 24n18, 199, 215n24, 229, 231, 232, 238n4, 239n11
Heisenberg, W. 179, 188n4
hermeneutics 229, 230
Hipparchus 97, 100
Husserl, E. 1, 191, 200–204, 210, 212n7, 214n17, 214n21, 226–229, 231, 232
  protention and retention in 201, 203, 210
Huxley, T.H. 153

immanence 58, 145, 158, 159, 168, 197, 198, 202, 207
impetus [*Élan*] 59–61, 63n46, 67, 68, 71, 73, 77, 83, 94n8, 99, 146, 161, 196, 212n7
indeterminacy 18, 37, 154, 157, 159, 180, 203, 209
inertia 103, 106–108, 178, 203
inevitability, feeling of 37, 52, 54, 92
infinitesimal 70, 84, 177–179, 209
infinity 2, 5, 6, 12, 32, 33, 115, 117, 125, 126, 128, 131, 137, 139n8, 154, 159, 163, 167, 169, 172, 178, 207, 210, 215n29

Index 245

of division (Aristotle) 5, 12–17, 21n2
inflationary model of universe 116, 129
instant Ch1 *passim*, 25, 33, 38, 40, 48, 50–56, Ch4 *passim*, 138, 145, 148, 149, 156, 167, 170, 177–179, 182, Ch9 *passim*
complex 1, 20, Ch9 *passim*
creative 64, 68, 170
dimensionless 5, 11–13
inseparability of contents of 52, 196
rhythm of 88–90, 197
strange nature of 10–12, 16, 24n16
intensity 40, 48–50, 57, 72, 78, 91, 154, 165, 169, 172, 194, 221
intentionality 89, 200–211, 232
interval 6, 26, 55, 61, 81, 85, 91, 148, 149, 205, 206
introspection 36, 96
intuition 26, 35, 67–74, 81, 82, 87, 90–92, 93n2, 98, 99, 171, 172, 224, 228, 229, 232

Jacob, F. 162
James, W. 62n21
Janet, P. 37–39, 42, 63n37, 214n16

K meson 184
Kant, I. 158, 175n21, 231–233, 236
Kauffman, S. 112, 114, 131, 137, 139n1
Kepler, J. 97

laminar flow 208
language 29, 51, 58, 59, 80–82, 85, 90–92, 100, 150, 151, 210, 219, 224, 227, 230–232, 234, 237
nouns and verbs 99–106
Leibniz, G.W. 24n17, 95n15, 101, 119, 158, 176n31, 227
Levinas, E. 20, Ch9 *passim*, 237

Il y a in 204, 215n22
life Ch3 *passim*, 65–74, 80, 82, 83, 86, 87, 94n4, 94n8, 95n21, Ch7 *passim*, 193, 194, 198, 199, 201–205, 209, 212n7, 232, 236
unrolling of psychological 48, 153
light 30, 33–35, 58, 106–108, 116, 128, 134, 155, 159, 160, 171, 186
Liouville equation 221
logos 3, 4
longitude 33, 34
loop quantum gravity 125, 132, 141n29
Lucretius 191, 207, 208, 210, 215n32, 215n34, 215n35

McAllester Jones, M. 213n10
McTaggart, J.M.E. 233
Mach, E. 101, 103
Mallarmé, S. 64, 81
manifold 12, 114, 119, 123, 132, 137, 139n7, 141n25
see also multiplicity
Markopoulou, F. 112, 114, 118, 128–130, 135, 136, 138, 139n1
Maxwell, J.C. 180
mechanism 44–46, 50–53, 60, Ch7 *passim*, 209, 210
memory 25, 64, 65, 72, 78, 86–89, 100, 104, 105, 149, 196, 228, 232
actualisation of 60
difference with perception 38–40, 46–50
as doubling perception 41, 51–54
formation of 46, 105
memory-image 41, 47
of present 52–57, 60
pure 47, 49, 51, 78
as trace in the brain 46–48, 105
Merleau-Ponty, M. 225
Michelson, A.A. 76

Minkowski, E. 197, 213n13
Moby Dick 163
motion/movement Ch1 *passim*, 23n14, 24n17, 27, 28, 75, 97–108, 127, 140n11, 177, 178, 184, 196, 199, 202, 208, 213n12, 215n26, 215n35, 217–221
multiplicity 77, 88, 146, 155, 196, 237
  continuous (virtual) 63n43, 145, 146, 151, 212n6
  discrete (actual) 148, 151, 157, 193
music 85, 88, 226–228

natural selection 131, 148, 149, 153, 154, 157, 159–163
neutrality, ontological (*epoché*) 226, 227, 230
Newman, E.T. 112, 114, 127, 136–138
newness/novelty 65, 73, 79, 80, 91, 114, 116, 130, 145, 149, 153, 157, 165, 172, 190, 192–198
Newton, I. 24n17, 29, 33, 101, 127, 178–180, 217, 218, 221
Nietzsche, F. 162
now Ch1 *passim*, 22n9, 23n13, 24n18, 76, 102–106, 177
  eternal 8, 228
  punctual 9, 22n9, 80, 102, 201, 225
  dividing/joining time (Aristotle) 18
  same/different (Aristotle) 16
nothingness 64, 70, 72, 73, 76, 77, 79–81, 90, 91, 155, 191, 204, 205, 207, 215n24–25

observables Ch6 *passim*
Owen, G.E.L. 5, 23n13, 24n17

Paley, W. 166

Parmenides 3–10, 12, 21n1, 22n4, 204, 207, 208
past 3, 4, 8, 14, 16, 17, 25, 30, 31, Ch3 *passim*, 64–69, 76, 80, 87–90, 104, 105, 116, 126, 129, 134, 135, 138, 145, 148, 149, 151, 156, 158, 165, Ch8 pasim, 195, 201, 203, 218, 234
  in general 37, 51
path integral formulation 126
perception 7, Ch3 *passim*, 97, 150, 155, 158, 214n20
perception-image 41
phenomenology 2, 145, 191, 199, 200, 202, 203, 207, 211, 217, 219, Ch11 *passim*
plane 169, 187
  see also present, plane of
Plato 1, 2, 6, 7, 9, 11, 12, 16, 17, 19, 23n13, 23n14, 24n17, 157, 175n21, 191, 228
  dilemma of participation in 6
  forms (ideas) in 6, 7, 11, 169
  Parmenides 6–9, 11, 22n3
  Timaeus 8, 23n14
poetics 21, 196, 198, 225, 226, 235, 237, 238
Poincaré, H. 1, 96, 101, 102, 220, 222
point 6, 9, 10, 16, 23n14, 33, 37, 69, 70, 75, 77, 85, 101, 102, 106, 114, 115, 117–121, 139n7, 151, 168, 169, 177–179, 197, 199, 201, 225
possible (and real) 72, 91, 102, Ch6 *passim*, Ch7 *passim*, 177, 182, 186, 188, 195, 206
potentiality 11, 13, 17–19, 24n15, 72, 88, 105, 149, 153, 180, 182, 190, 213n7, 214n18, 219
present 25, 30, 31, Ch3 *passim*, 64, 65, 67, 68, 75, 87–90, 145, 148,

Index

149, 151, 156, Ch8 *passim*, 195, 200–203, 205, 207, 234
continuity of 145, 177
continuous encroachment on future 58, 156
falling back towards the past 48, 165
fractal 1, 21, 187
eternal 228
memory of 52–57, 66
plane of 1, 21, 177–180, 187
splitting of 1, 24n22, 47, 51, 53
springing forward of towards the future 48, 213n12
upsurge of 48
Prigogine, I. 176n41, 217–219
process philosophy 99, 100, 105
Ptolemy 97

quantum cosmology 19, 112, 113, 122, 130, 134–138
quantum gravity 19, 113, 125, 132, 135, 141n29
quantum mechanics 1, 19, 117, 123, 179–183, 217, 221
  Copenhagen interpretation 20, 181, 183
  interpretational paradoxes 181
  transactional interpretation 21, 181–183

Rapaport, H. 241n29
relation 7, 8, 14, 16, 31, 57, 88, 101, 102, 118, 119, 128–130, 145, 152–156, 164, 167, 168, 171, 174n11, 175n21, 176n30, 179, Ch9 *passim*, 227, 229, 234, 238n7
relativity 1, 75, 80, 92n1, 101, 102, 130, 150, 174n11, 181, 213n10
  general theory of 1, 19, Ch6 *passim*, 179

special theory of 1, 101, 106–108, 179, 181, 188n5
repetition 40, 43, 50, 54, 71, 85, 86, 145, 153, 182, 228
rhythm 79, 82–84, 88–90, 95n16, 151, 154, 194–197, 211, 227, 228, 229, 237
Ricoeur, P. 231–237
Roupnel, G. Ch4 *passim*
Rovelli, C. 114

Schrödinger, E. 180, 188n4
Schrödinger equation 124, 180–182, 221
science 2, 26, 75, 91, 92, 92n1, 93n2, 95n13, 96–99, 103, 134, 144, 148–151, 155, 157, 167, 169, 172, 173, 176n41, 190, 207
  metaphysics of 149, 150, 172, 175n21
  method of 96–99, 103, 104, 110n6
sensibility 172, 190, 196, 200, 214n20
Serres, M. 20, 190, 191, 207–209
Shimony, Abner 100
sidereal clock 27
simultaneity Ch2 *passim*, 40, 76, 101, 136, 177, 179, 202
Smolin, L. 20
Solvay Institute 91, 95n20
space 1, 6, 12, 26, 72–75, 80, 89, 97, 98, 100–102, Ch6 *passim*, 141n29, 147, 151, 162–169, 178, 181, 208, 209, 217–222, 225, 226, 229, 237, 239n11
  Hilbert space 123, 130, 132–134
spin 132–137, 181, 187
Spinoza, B. 77
splitting of personality 41
sugar 49, 74, 75, 94n8, 152, 174n13
supergravity 119
supersymmetry 125, 133
synchronism 82–85, 197

tendency 20, 38, Ch7 *passim*, 218, 219
tension (tone) of psychological life 39, 42, 43, 60, 63n41, 89
thermodynamics, second law of 104, 184, 218–220
thermodynamic arrow 184–186
Thompson, W. 218
thrownness 230
time
 absolute 72, 74–76, 101, 102, 104, 120, 178, 179
 arithmetisation of 74, 80, 83, 84
 arrow of 103, 104, 156, 183–186, 189n13, 218, 221
 atomisation of 74, 193
 being in 8, 9, 11
 capsule 1, 19, 105
 and causality 2, 68, 92, 126, 135, 138, 186, 194–196, 206, 210, 214n18, 232, 236
 continuous 1–4, 8, 13–15, 19, 21n2, 66, 69, 76, 82–84, 92, 177, 188n1, 190, 192–196, 199, 206, 209, 230, 232, 234
 diachrony of 203, 205–207, 210
 and determinism 154, 155, 178, 179, 183, 186, 187
 direction of 87, 103, 104, 182, 183, 185, 217
 discontinuous 19, 21, 23n13, 25, Ch4 *passim*, 149, 190–194, 205, 206, 227
 economy of 226, 227, 229, 230, 236, 237
 elimination of 112–127, 134, 135, 149
 existence of 80, 81
 fluid nature of 99
 no intuition of simultaneity 35
 irreversibility of 217, 220–222
 lapse of 75
 as line 9, 16, 68, 69, 72, 77, 85, 104, 151, 156, 177, 196, 199, 207, 229, 234
 measurement of 14, 17, Ch2 *passim*, 72–78, 81–86, 101–103, 105, 106, 115, 117–121, 136, 188n1, 193, 201, 209
 nature of 2, 8, 11, 14, 16, 74, 80, 177
 as number of motion 15, 18, 231
 openness of 18, 157
 parameter 10, 118–126, 217
 passing of Ch1 *passim*, 23n14–16, 64, 65, 75, 81, 87, 98, 156, 191, 192, 217, 228
 poetics of 21, 196, 198, 225, 226
 psychological 25, 26, 30, 78, 85, 107, 145, 228
 qualitative/quantitative 74
 reversibility of laws of physics 217–222
 shelters 1, 21, 225–231, 235
 snapshots of 99–106, 151
 space-time 75, 80, 82, 102, 106–108, 114, 115, 126, 127, 129, 137, 139n8, 177, 179, 182, 183, 219
 symmetry, breaking of 1, 184, 217, 220, 221
touch 9, 51, 206–211
transcendence 7, 11, 17, 145, 146, 152, 161–163, 169, 173, 200, 202, 206, 207, 210, 214n20, 226
turbulence 208
twin paradox 106–109

uncertainty principle 179, 180
universe 2, 19, 20, 32, 91, 96–104, 106, Ch6 *passim*, 144, 152, 153, 155, 157, 167, 170, 171, 172, 177–179, 184, 186, 188, 217, 218, 226, 230
 clockwork 178, 180
 possible 1, 121, 138

Index

utility 56, 150, 151, 154, 155, 171

virtual 20, 51, 53, 57, 91, Ch7 *passim*, 195, 202
 and actual 51, 144–147, 168, 169
 and actual as two aspects of instant 51
vis viva 29, 35n3
vital impetus *see* impetus
vitalism 171
void 5, 71, 73, 145, 191–194, 204–209, 215n25, 215n34–35

wave, advanced and retarded 180–183, 186
wave function 123, 124, 134, 180–183, 188n1
Webb, D. 20
Wheeler, J. 102, 122, 181, 182
Whitehead, A.N. 99, 106
whole 7, 8, 14, 52, 58, 67, 79, 117, 120, 129, 153, 167–170, 206, 208, 226
will 36, 39, 46, 53, 61, 69, 72, 78, 79, 161, 188, 196, 233
Wittgenstein, L. 191
Wood, D. 3, 21
 Deconstruction of Time 240n12